DESIGNING WITH
TILE, STONE & BRICK

The Creative Touch

Carol Soucek King, Ph.D.

Foreword by Stanley Abercrombie, FAIA

Interior Details

AN IMPRINT OF

PBC INTERNATIONAL, INC.

Distributor to the book trade in the United States and Canada
Rizzoli International Publications Inc.
300 Park Avenue South
New York, NY 10010

Distributor to the art trade in the United States and Canada
PBC International, Inc.
One School Street
Glen Cove, NY 11542

Distributor throughout the rest of the world
Hearst Books International
1350 Avenue of the Americas
New York, NY 10019

Library of Congress Cataloging–in–Publication Data

King, Carol Soucek.
Designing with tile, stone & brick: the creative touch / by Carol Soucek King.
    p.   cm.
  Includes index.
  ISBN 0–86636–328–9   (pbk ISBN 0-86636-398-X)
  1. Tiles in interior decoration. 2. Stone in interior decoration.
  3. Brick in interior decoration. 4. Designing with tile, stone, and brick. I. Title
NK2115.5.T54K56 1995   94–32129
728—dc20   CIP

CAVEAT– Information in this text is believed accurate, and will pose no
problem for the student or casual reader. However, the author was often
constrained by information contained in signed release forms, information
that could have been in error or not included at all. Any misinformation
(or lack of information) is the result of failure in these attestations. The
author has done whatever is possible to insure accuracy.

Color separation, printing and binding by
Toppan Printing Co. (H.K.) Ltd. Hong Kong
Design by Anistatia R. Miller

10 9 8 7 6 5 4 3 2 1

Printed in Hong Kong

To Creativity at Home ...

and Being at Home with Creativity!

CONTENTS

FOREWORD

Stanley Abercrombie, FAIA
Chief Editor, **Interior Design**

Sorry to repeat myself, but in the last sentence of a book titled *A Philosophy of Interior Design*, I wrote that interiors constitute "our most personal art." Carol Soucek King, I'm glad to see, seems to share the same view, for the admirable series of books Dr. King has planned promises to focus on just those aspects of interior design that make it personal.

The grand concept is not to be neglected, of course. Like any other art, interior design depends for its success on the encompassing vision that relates its many elements in a meaningful whole. But such vision, in interiors, becomes manifest and comprehensible through the medium of myriad details with which we are in intimate contact: the feel of a drawer-pull, the profile of a cornice, the polish and grain of wood, the "hand" of fabric.

This contact involves all our senses. We see our interiors, certainly, but we also smell the materials in them, we hear their acoustic properties, we brush up against their walls, step on their floors, open their casegoods, sit on their chairs. More than any other, interior design is the art we use. In that sense, it is not only our most personal art, but also the one most responsible for our well-being. In the context of increasingly brutalized urban environments, this is increasingly true and increasingly important. Interior design is often our refuge.

It is therefore a very welcome prospect that Dr. King is turning her experienced editorial eye to the details and materials on which the art of interior design depends. I'm sure we will all benefit from her discoveries.

PREFACE

Welcome to the world of tile, stone and brick as used by some of the world's leading designers to define and empower the spaces in which we live. The vast array of sizes, textures, colors and patterns available today enables these age-old materials to inspire creative ideas as well as to help their users turn those ideas into reality.

From the front gate to the back door and everywhere in between, uncut stone, cut stone, slate, granite, marble, ceramic, terra cotta, newly introduced composites and age-old alternatives...the list continues!...are making the design of one's home a more exciting experience than ever before.

Given the task of searching the world for some of the best uses of tile, stone and brick in homes today, I asked more than fifty architects and interior designers throughout the world to share their recent projects. The result is a collection of styles and applications as diverse as the materials themselves.

Through these projects we can see how new color and design trends, new techniques and new technologies available through tile, stone and brick can be used to create better homes for better living. More important, these designers show how, whether in a newly built home, a remodel, a high-rise apartment or a "home away from home" vacation spot, the beauty and quality available through tile, stone and brick can be part of everyone's daily life.

Carol Soucek King, Ph.D.

INTRODUCTION

Tile

Peter C. Johnson, Jr.
Vice Chairman of the Board, **Summitville Tiles, Inc.**
1994 Chairman of the Board of Governors, **International Tile & Stone Exposition**
Past Chairman, **Tile Promotion Board**
Past President, **Tile Council of America**

For thousands of years, ceramic tile, natural stones and brick have been the surfacing choice for royalty, the rich, the church and the government. Throughout the world you can see beautiful installations of yesteryear still intact in cathedrals, monasteries, castles and government buildings.

Today, with the improvements in manufacturing and installation, what was once only affordable to the elite is now available and affordable to the average working citizen.

There was a time not long ago when tile could only be mud-set in a thick mortar bed. Wall tiles were soaked in fifty-five-gallon drums for twenty-four hours before installation. Special construction techniques had to be adhered to so buildings could accept tile. Masonry buildings were more tile-friendly than buildings of wood frame construction. Thus, the cost of ceramic tile in the home was very expensive.

Today tile can be set directly over a multitude of surfaces — including tile over tile. The thin-set method of tile installation has literally revolutionized the hard surface industries of ceramic tile and natural stone. "Do-it-yourself" installations are quite common for the handyman of today.

It wasn't that long ago when clay was dug by hand, crushed by stone grinding wheels, mixed with water and hand formed, dried and fired for a couple of weeks in a crude kiln fired by wood or coal. Decorative tiles were made by artisans and were quite expensive. Today, raw materials are mined by huge earth-moving machinery before being crushed and atomized in state-of-the-art spray driers. Computer-driven production lines and robotic equipment have replaced the artisans of yesteryear in many factories today, and firing time has been reduced from a couple of weeks to under an hour.

The result has been a virtual explosion of new techniques, new sizes, new colors, new textures and new ideas — all at affordable prices.

It is especially important for any consumer to obtain the proper installation instructions from a qualified expert before starting a weekend tile job. Some jobs are still best left to the professionals. It is very important to always follow recognized industry standards for the installation and selection of ceramic tile, natural stones and brick. What may perform well in a home may not work around a pool deck. Glaze wear, freeze-thaw resistance, acid resistance and stain resistance are a few examples of physical properties consumers need to be aware of. When in doubt, check your yellow pages and get recommendations in writing.

Dr. King has illustrated just a few of the millions of beautiful installations of natural stone, ceramic tile, brick and related products in this book. A properly installed product will add value to your home and provide you and your heirs with years of aesthetic beauty with minimal upkeep. With all of the great advantages offered by natural stone, ceramic tile and brick, I hope this book inspires you to turn your home into a castle.

Stone

Robert Hund

Managing Director,
Marble Institute of America

How did man discover the beauty hidden in stone? Probably curiosity. Someone saw unusual markings on an exposed ledge of rock. Breaking a piece off, smoothing it and cleaning it, rubbing it with pumice and wetting it revealed colors, veining, texture and graining. Obviously, the person who could transform raw materials into valuable assets could become an advisor to the king, if not the king himself.

How long has man used stone to symbolize permanence? Who knows? To work with stone is to work with the basic rhythms of the earth. We do know that by the time the pyramids were built, man had learned how to work with stone, to shape and place it according to his needs. The need for building materials in which beauty and permanence are prerequisites is greater now than ever before. To meet this increasing demand (a new "stone age," so to speak), machines that can cut and polish stone more precisely, thinner and faster than ever before have been produced. New materials which allow thick or thin stone to be installed at less cost have been developed.

This book shows how stone can be used. All it takes is the imagination of the designer drawing upon the world's resources. And best of all, as the earth ages, it renews itself in terms of stone resources.

In this age of environmental awareness, brick is also being recognized as a building material that is environmentally friendly throughout its life. Clay raw materials are abundant and efficiently used. Brick contributes to the strength, durability and energy efficiency of the buildings in which it is used. Brick buildings are an important part of passive solar structures. Because of brick's mass, lower amounts of insulation may be required. The longevity of brick masonry structures is difficult to overlook.

Brick

Brian E. Trimble

Senior Engineer, **Technical Services**
Engineering and Research Division
Brick Institute of America

Brick has been recognized for thousands of years as a sturdy, attractive building material that can provide shelter from natural elements. Brick provides a multifunctional purpose in homes and buildings. Applications include walls, pavements, columns, piers, retaining walls, screen walls, planters, fireplaces, serpentine walls, soffits and beams. Brick walls not only serve as enclosure and aesthetic enhancements, but also act as structure. Brick masonry walls also serve as a barrier to fire.

Brick offers designers endless design possibilities, with many colors, textures, patterns and shapes. Using combinations of these leads to truly dramatic architecture. World-renowned architecture has been achieved with brick — the Colosseum, the Great Wall of China, Westminster Abbey, Monticello, and even Camden Yards Ballpark. No material offers all of these choices in one building material.

Although brick has been around for centuries, new innovations in the use of brick continue to occur. Stronger masonry structures have been developed that rise to new heights. Prefabricated brickwork has allowed new ways to construct buildings. And mortarless systems in pavements and retaining walls have allowed brick to be constructed more economically than in the past.

URBAN

SOPHISTICATION

MAJESTIC IN MEXICO
Limestone, Granite, Marble, Kirkstone, Recinto & Slate

■ ■ ■

THE VAST VARIETY OF STONE ONE FINDS in the projects by **Gerard and Carlos Pascal** is one of the reasons the work produced by Pascal Arquitectos is so powerfully connected to the land.

"Usually, when we think about architecture, we think about stone," says Gerard Pascal. "Maybe it's because that's how it started. Man's first habitat was stone, and stone was the most available material then and still is today. Also, in some countries, such as here in Mexico where we live, stone is still used in the same traditional ways, linking us to the past, to memories we don't want to lose."

ABOVE: *A majestic fireplace carved from Fontainebleau limestone, France.*

LEFT: *Completing a circle created by leather sofas from J. Robert Scott & Associates, a fireplace of Chiluca limestone from Mexico.*

OPPOSITE: *Composed of Huixquilucan stone, the volcanic rock recinto, and black St. Gabriel granite, this fireplace provides a commanding niche for a Batak shield from Polynesia.*

Photography by Victor Benitez

LEFT: *Light plays across Teca limestone walls and a marble Byzantine mosaic designed by Pascal Arquitectos, realized by Luis Scodellere.*

BELOW: *For a dramatic foyer, the consistent muted tone of Luxor limestone from Mexico is juxtaposed with gray and pink Del Salto granite from Argentina and black Tijuca granite from Brazil.*

OPPOSITE: *Pascal Arquitectos' sculptured staircase and handrails rise from a black slate floor surrounding a fountain of pink Labrador granite from Norway and black St. Gabriel granite from Brazil.*

Photography by
Victor Benitez (opposite and left)
and Eitan Feinholz (below)

ABOVE: *A floor of Indian Paradise granite is inlaid with white Thassos marble and Black Andes granite.*

RIGHT: *Daylight pours in through a skylight of white alabaster from Spain. The wall is of Huixquilucan stone, Mexico. The floor is Indian Paradise granite.*

Photography by Eitan Feinholz (above) and Victor Benitez (right)

LEFT: *Reflected in Pascal Arquitectos' custom-designed mirror are a Black Andes granite countertop and Silk Dynasty wallpaper.*

BELOW: *This work of art for the floor is composed of green Labrador granite, white Thassos marble, and Black Andes granite.*

Photography by Victor Benitez

OPPOSITE: *Standing out in relief against a façade of Huixquilucan stone are a fountain and its surrounding paving of recinto (volcanic) stone.*

RIGHT: *Creating a quiet corner of a garden is a fountain of sandblasted Kirkstone, a metamorphosed volcanic ash quarried in England, and Baltic Brown granite.*

BELOW: *A fountain wall of pink Huixquilucan stone from Mexico rises from Nambam flamed granite from Australia.*

Photography by
Victor Benitez (opposite)
Eitan Feinholz (right and below)

DECISIVE MINIMALISM
Granite, Marble & Ceramic Tile

■ ■ ■

FUNCTION, ORDER AND CRAFTSMANSHIP — these qualities exude a commanding presence in all the work of the Chicago-based interior architecture firm, Powell/Kleinschmidt. While **Donald D. Powell** concentrates on planning and **Robert D. Kleinschmidt** on design, the partners typically collaborate on all projects to produce minimal modernist work defined by a recognizable confidence in knowing what is necessary... and what is not. In this context it is immensely rewarding to the mind to see their decisive strokes in dealing with

ABOVE: *In redesigning a penthouse in a Chicago high-rise, Powell/Kleinschmidt left unchanged this original 1957 bathroom by the building's architect, Mies van der Rohe. Restoration by Midwest Marble & Granite Co.*

Photography by Jon Miller/Hedrich-Blessing

RIGHT AND OPPOSITE: *Mies van der Rohe designed this Chicago penthouse for the building's developer and his greatest patron, Herbert Greenwald. Repeating the building's aluminum exterior, the staircase connecting living and sleeping levels is made of aluminum with standard white Vermont marble treads, with honed finish. "It's a piece of sculpture in space," says Powell, who elected to do nothing to change the original design. The marble stairs and floor, originally installed by Peerling-Sheddy Marble Co., were cleaned with pumice and restored by Midwest Marble & Granite Co.*

RIGHT: *Furniture as small-scale architectural elements and necessarily just as functional is a constant in the work of Powell/Kleinschmidt. Here in the dining area of an apartment in another Chicago high-rise designed by Mies van der Rohe, the dining table's three-foot-by-eight-foot granite top slides out from the banquette twelve inches on hidden glides to allow guests easy access. A honed rather than highly polished finish minimizes glare. Fabricator was Midwest Marble & Granite Co.*

OPPOSITE AND BELOW: *For another apartment, Powell/Kleinschmidt associate William Arnold maintained a consistency of materials throughout by selecting Impala Black granite for counter and tabletops, bleached ash for the flooring, and American cherry for the cabinetry. To make the dining table appear to be one thick slab of granite, yet in fact be light enough to be expandable, the designers used one-fourth-inch-thick granite backed to a honeycombed aluminum grid.*

Photography by William Kildow (opposite and below) and Jon Miller/Hedrich-Blessing (above)

the materials in their always-limited
palette.

The extreme clarity of planes seen in
their work was a strong concept in the philosophy of Ludwig Mies van der Rohe,
whose influence in the work of
Powell/Kleinschmidt is major and
in whose architecture many of
Powell/Kleinschmidt's residential
projects have been completed.

So Continental in SoHo

Granite, Limestone & Tumbled Marble

■ ■ ■

ONE MIGHT NEVER EXPECT TO FIND such a Continental look in the SoHo section of New York City, but that is the feeling **Penny Drue Baird** created in a completely raw, four-thousand-square-foot loft. Following her design philosophy, which is to always use authentic rather than synthetic materials, the founder of Dessins Inc. set her mix of French antiques and modern SoHo architectural components against a variety of marble, both slab and tumbled, and granite.

ABOVE: *Generously sized squares of polished granite provide a fitting prelude to the luxurious master bath. Stone throughout is from Country Floors.*

LEFT: *Slab marble, used for the bedroom fireplace surround, is in harmony with the designer's collection of French antiques.*

OPPOSITE: *In the colonnade, a rustic, tumbled-marble floor provides adjacent rooms with a decidedly country French connection.*

Photography by Jon Elliott

SÃO PAULO ACCENTS
Marble, Granite & Ceramic Tile

■ ■ ■

BRAZILIAN DESIGNER *Arthur de Mattos Casas'* current work emphasizes wood, with stone used primarily as an accent. Yet accents in stone by Casas, such as this marble fireplace in one area of his São Paulo home, are usually crescendos!

Being able to create striking patterns within his typically limited color palette is one reason Casas selected ceramic tile for his home's master bathroom. Ceramic's durable and cleanable quality is another.

BELOW: *For this fireplace, Arthur de Mattos Casas used Spanish black and Italian green marble from Itapemirin to create a dialogue with the home's 1940s modern architecture.*

Photography by Tuca Reinés

ABOVE LEFT AND RIGHT:
*In his master bath, Casas
uses HGK black and white
ceramic tile, placed on an
estuque of metal, cement
and plaster executed by
artist Herman Tacasey.*

LEFT: *"Polished nobility"
is the quality Casas likes
about the black granite
from Itapemirin he selected
to contrast with the estuque
by artist Herman Tacasey.*

LEFT: *Roche d'Harveaux limestone, cut into sixteen-inch squares and placed in a running bond pattern, is used in all remodeled areas. The limestone picks up the tone of the maple cabinetry, while the black granite, devoid of grain, provides contrast. In the splash area, all electrical receptacles and switches are specifically fabricated from the granite. Absolute Black Zimbabwe granite and Roche d'Harveaux limestone were supplied and fabricated by Studio Marble, Inc., except for the granite electrical plates made by Di Camillo Marble Accessories.*

BEVERLY HILLS HOSPITALITY
Granite & Limestone

■ ■ ■

ARCHITECT *Alison Wright* HAS USED much granite and limestone in these remodeled kitchen and bath areas for a client who entertains frequently. Since the client, Peter Preis of Beverly Hills, had discovered that today's guests always seem to congregate in the kitchen, Wright, along with interior designer *Irene Montgomery* and kitchen consultant *Chris Tosdevin* of Bulthaup, tailored it to look like a library. Maple cabinetry, discreet appliances, Absolute Black Zimbabwe granite counters and splash, and Roche d'Harveaux limestone flooring emphasize sleek sophistication rather than utility.

The limestone continues in the guest bath, exercise room and exterior deck areas for a sense of connection that is further emphasized by the use of the granite elsewhere — for steam room recesses and within the shower. Throughout, the designers used stone not as applied decoration, but as a building material. The result is impressively strong as well as elegantly hospitable.

OPPOSITE: *The designers complemented the guest bath walls of twelve-inch Roche d'Harveaux limestone squares with recesses of the Absolute Black Zimbabwe granite and a terra-cotta sink — a combination that evokes the mystique of an exotic Turkish steam room. Contractor for all new work in this residence was Dominque Rocoffort, ROC Development.*

Photography by Jerome Adamstein

DISTILLED PRECISION
Absolute Black Granite & Marble

■ ■ ■

THE SERENITY OF POLISHED ABSOLUTE Black granite and the purity of white sculpted marble—these are major elements in the concisely edited and dramatic interiors of **Sally Sirkin Lewis.** They counterbalance the lush textures of the textiles she uses and offset her wood veneers and frequent exclamation points of gold and silver leaf. They also provide contrast in her always minimalist color palette.

LEFT: *In her own Beverly Hills residence, Sally Sirkin Lewis leans toward the architectonic quality of stone...even in the accessories, as shown by the limestone Roman bust highlighting the media unit.*

RIGHT: *Absolute Black granite serves as counterpoint to the blond silk walls; its thirty-inch grid conforms to the width of adjacent veneered wall panels.*

OPPOSITE: *In the New York showroom of J. Robert Scott & Associates, of which she is president and designer, Sally Sirkin Lewis uses stone to continue her highly edited play of contrasts. Here, an Art Deco limestone sculpture by Henri Ernest Parayre stands on a faux marble "Origami" console.*

Photography by Jaime Ardiles-Arce (above and opposite) and Alex Vertikoff (right)

OASIS IN RIYADH
Granite, Limestone & Marble

■ ■ ■

JUST AS **Erika Brunson** HAS BECOME known for transporting some of the finest European furnishings to California, she has also become known for transporting her highly civilized approach to living in the Sunbelt to balmy climes elsewhere. In fact, for this residence in Riyadh, she selected and made all the furniture in Los Angeles, then shipped it to Saudi Arabia. Since the scope of the assignment, which she carried out with interior architect **Anthony Machado**, was to completely furnish **Theodore Ceraldi's** architectural design covering eighty-five-thousand square feet, that shipment consisted of nineteen forty-foot containers!

Photography by Billy Cunningham

OPPOSITE AND ABOVE: *Stairs of Absolute Black granite wrap around the split-faced Golden Sierra travertine wall. All stone and marble are from Carrara, Italy, except for the Absolute Black granite, which was mined in Saudi Arabia, shipped in blocks to Carrara where it was worked, and then shipped back to Riyadh.*

OPPOSITE: *Wicker Works chairs and Country Floors custom-colored ceramic tiles, imported to the United States from Portugal, stand out against a glorious backdrop — a fountain designed by Erika Brunson and made of honed Botticino marble. Water runs over the pillars and into the pool.*

RIGHT: *The strength of the stone walls and floors is carried through in other furnishings, including stone chairs by Michael Taylor Designs and vast murals by Douglas Riseborough. The vertical panel consists of chevron-patterned sandblasted glass with a painted grillwork screen.*

BELOW: *One of two reception rooms featuring vaulted ceilings and bisected by a wall of split-faced Golden Sierra travertine. The floor combines Anthennia Verde, Botticino and Bianco Scaglia marbles, polished limestone and Absolute Black granite.*

Photography by
Billy Cunningham

Seaside Retreats

SEA ISLAND REVERIE
Beach Stones

■ ■ ■

THE BEACH HOUSE WHICH ARCHITECT *John Portman* built for his family on Sea Island, Georgia, is similar to the dramatic atrium hotels he has designed in terms of its monumental expanses of smooth white concrete and glass, explorations of space, form and light. However, also as in his hotels, amid these sweeping geometries he does create oases of a more earthy sort, often through the use of stone. Here, his driveway of granite-colored concrete forms a grid for a carpet of green, and the large stones in the reflecting pools in the interior courtyard pay homage to nature's unending individuality.

OPPOSITE AND ABOVE: *In the reflecting pool, beach stones are a larger-than-usual size to complement the magnitude of the space and provide another link to the natural environment at Entelechy II. The sculpture, "Reverence," is by Arno Breker.*

RIGHT: *For the driveway to John Portman's Entelechy II beach house, a geometric concrete grid creates a base that was filled with soil and seeded to create a surface that is both durable and lush.*

Photography by Michael Portman

MAUI MAGIC
Limestone, Flagstone & Marble

■ ■ ■

ARCHITECT ***Charles Lau***, PRINCIPAL of AM Partners in Honolulu, and interior designer ***Lois Zanteson*** had much to inspire them when creating this residence in Maui. The new signature home for Kapalua Land Company's Plantation Estates, the site is adjacent to a PGA championship golf course and offers incomparable views across the links to the ocean beyond. When the client asked for architecture, interiors and landscaping consistent with such natural surroundings, they used stone throughout as a key element in achieving the desired unity. The formal areas of the house feature a Portuguese limestone floor bordered by a slightly deeper tone of the same earthy beige. Limestone slabs are used on the custom-designed fireplaces and grand staircase. The bathrooms combine the soft, matte texture of limestone with the soft, elegant sheen of marble. For the pool deck and lanais, natural paving blends with the magnificent scenery. Random-cut Arizona sandstone was selected for the broad curving forms which are inset with colorful flower beds and bordered by low stone walls.

The effect is a sense of relaxed formality so suitable for both the plantation theme and the surroundings that, rather than the reverse, one might wonder if it wasn't the home that inspired Mother Nature!

ABOVE: *In the bathrooms, marble from Walker/Zanger joins limestone for a heightened sense of elegance.*

BELOW: *The Classic Oak color of random-cut Arizona sandstone supplied by Denny Moore, inset with flower beds and bordered by low stone walls, provides a durable yet gracefully-shaped surface for the pool deck and lanais.*

OPPOSITE: *Throughout the home, formal areas are linked by Portuguese limestone floors from Walker/Zanger.*

Photography by Klein & Wilson

TRADITIONAL IN MEXICO
Conchuela, River Stones & Marble

■ ■ ■

THE WORK OF GOMEZ VAZQUEZ ALDANA & Associates always uses a multitude of native materials to reflect local folklore and emphasize traditions deeply rooted in Mexican artistry. An additional benefit is that using local sources helps to keep budgets within limits, even though the luxurious look may appear otherwise. This most certainly was the case in two of the firm's recent projects directed by architects *Jaime* and *J. Manuel Gomez Vazquez*, in association with interior designer *Marisabel Gomez de Morales*.

ABOVE AND OPPOSITE: *Located on Cancun's waterfront, Casa del Sol consists of five separate villas for a client who wanted to feel contemporary, comfortable, and bathed in unabashed luxury. In addition to the high, sloped, wood-beamed ceilings, overstuffed sofas and expansive windows, many stones native to Mexico were incorporated. These include piedra bola (round river stones) used in bold, intricate patterns across a floor of Blanco Guadiana, Verde Tikal and* *travertine marble from Italy. Walls were treated with conchuela stone, which is a natural fossilized limestone with shell imprints. Piedra bola was supplied by Terrajal. Conchuela stone was supplied by Camasa. Marble was supplied by Marmol y Arte.*

Photography by Scott McDonald/Hedrich-Blessing

The warm organic colors, wood and irregular stone flooring contribute to architect Steven Erhlich's expression of California's indoor/outdoor lifestyle. Idaho flagstone was supplied by Ray Stutzke and installed by stone mason Eugene Salerno.

Photography by Grey Crawford

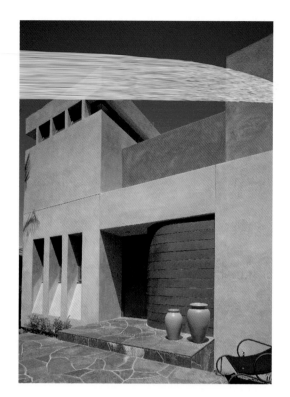

CALIFORNIA SIMPLICITY
Idaho Flagstone

THERE IS A SEAMLESS QUALITY TO EVERY square inch, inside and out, of Hannah Hempstead's residence in the beach community of Venice, California. Architect **Steven Ehrlich**'s selection of a few earthy colors and a minimal number of materials to define simple, bold spaces and planes fuses façade, pool and garden with the rooms within.   Especially through the burnt sienna and yellow ocher of the stucco and the irregular flagstone flooring from Idaho, pulled off the side of Rattle Mountain near the Idaho/Montana border, he has turned his vision for a house in harmony with the natural environment into a unified, totally organic reality.

FLORIDA FANTASIES
Florida Keystone, Syenite & Stone Mosaics

■ ■ ■

THE CLIENTS, ENTERTAINERS WHO ARE often on tour, wanted a place in Florida where they could get away from it all, so architect **Ramon Pacheco** and interior designer **Nury Feria** created the feeling of an island retreat. In their elegant yet warm play of materials, Florida Keystone, cut coral which is also known as coquina, a form of limestone, takes the leading role. Combined with accents of Blue Bahia,

BELOW: *The undulating bar is surfaced with an impermeable Blue Bahia supplied by Empire Marble & Granite.*

ABOVE: *An obsolete exterior balcony was turned into an art gallery encompassed by modular cuts of Florida Keystone with glass shelves inserted into grooves for display.*

OPPOSITE: *Columns of the indigenous Florida Keystone define a pleasant alfresco setting featuring an artful, entirely custom display of handmade ceramic mosaic tile for both tabletop and swimming pool.*

Photography by Dan Forer

ABOVE: *Used as a constant backdrop in this play of natural materials, Florida Keystone from Keystone Products surrounds the fireplace and faces adjacent walls.*

OPPOSITE: *In her own home, Nury Feria worked with Ramon Cortina to mix three different stones — Rojo Alicante, Verde Oriental and travertine — for her stain-and-heat-resistant dining table.*

Photography by
Dan Forer (above) and
Barry J. Grossman (opposite)

a syenite from Brazil, handmade ceramic tile murals, rich woods and inventive uses of glass, its earthy texture sets the stage for this contemporary rendition of a traditional theme from beginning to end.

"The use of stone reflects an underlying concern for design to have a timeless quality," says Feria, who has worked magic with stone in her own home as well.

OCEAN VIEW IN EAST HAMPTON

French Limestone, Travertine & Flagstone

■ ■ ■

THE PERFORMANCE OF STONE IN HOT, humid weather is one reason **Alfredo De Vido** used it expansively here. The way it looks — fantastic — is another. Located near the ocean in East Hampton, New York, and designed for outdoor summer activities such as swimming, tennis and relaxed entertaining, the house makes use of various stone materials inside and out. French limestone, travertine marble and flagstone — all were chosen for their ability to endure all that wear and the moisture, too, and still maintain their original beauty and color.

BELOW: *Arizona flagstone from Southampton Brick and Tile lends a tan-red hue to terraces and poolside areas.*

BOTTOM: *The fireplace is a sculpture of flamed French limestone supplied by Hastings.*

Photography by Norman McGrath

The lofty entry begins the interior expanses of Italian travertine with the natural fissures filled with color-matched grout, furnished by Hastings Tile Co. The rug tapestry was made under license from artist Roy Lichtenstein.

OCEANFRONT TRANQUILITY
Marble & Granite

■ ■ ■

Louis Shuster OF FORT LAUDERDALE, Florida, is known for the way his pared-down interiors make already large spaces look endless. In this remodeled oceanfront condominium, full slabs and twenty-inch honed-and-filled squares of honey-beige Navona Travertine, sleekly polished black granite, and a neutral color scheme create a sense of tranquility that seems to extend beyond the horizon.

OPPOSITE ABOVE: *Full slabs of Navona Travertine frame the elevator doors and form a pair of floor-to-ceiling columns which flank and support the lobby shelf, making it appear to float. Stone throughout was supplied by Intercontinental Marble and installed by Village Floors.*

OPPOSITE BELOW: *The dramatic contrast established by the light travertine and dark granite is echoed in all other furnishings.*

ABOVE: *On walls, hand-painted paper mimics the feeling underfoot of polished black granite and travertine.*

Photography by Kim Sargent

Italian Holiday
Ceramic Tile

■ ■ ■

THE WORK OF STUDIO TRANSIT DESIGN'S *Giovanni Ascarelli*, *Maurizio Macciocchi*, *Evaristo Nicolao* and *Danilo Parisio* often revolves around one of the firm's primary beliefs — that public spaces should make people feel as good as if they were at home. Here, tile was used not only for its long life and easy maintenance, but also for its ability to provide clarity of form and color — without which vast public areas can easily lose the very sense of life they are intended to serve.

At the Grand Hotel Masseria S. Lucia, brilliant blue ceramic tile emphasizes the hues of sky and water and defines the shape of the pool. Tile was supplied by Asper Studio.

Photography by Janos Grapow

COUNTRY
REVISITED

EARTH-SHELTERED IN CONNECTICUT
Fieldstone, Bluestone & Slate

■ ■ ■

AN EARTH-SHELTERED HOUSE, THE home of Richard and Noriko Moore in northern Connecticut uses timbers cut from the site for its posts and beams, as well as fieldstone from the site for facing on reinforced-concrete retaining walls, and on free-standing site and garden walls. Buckingham slate from Virginia and blue-stone slabs from Pennsylvania are used to surface concrete floors as a heat-retaining measure. Solar heat is gained through the south-facing glass and absorbed by the stone-concrete combination. It is then stored and reradiated at night or in cool periods.

The architect, **Alfredo De Vido**, who worked in collaboration with his client Richard Moore, says, "stone, as an attractive, durable and dense material, is quite suitable for such an earth-sheltered house, in which one of our main goals was to use features that would conserve energy."

ABOVE: *Virginia slate and Pennsylvania bluestone make use of the sun's energy through their ability to absorb and retain heat.*

OPPOSITE: *Natural light is brought into the rear of the house via a long row of skylights, emphasizing the rich colorings of wood and stone and providing a light and airy feeling even on overcast days.*

RIGHT: *The house was sited so as to merge into its natural environment with minimal disturbance to the land. Some of the materials such as the fieldstone, found on the site itself, make the house at times seem indistinguishable from the woods beyond.*

Photography by Norman McGrath

LIVING HISTORY
IN PENNSYLVANIA
Fieldstone, Flagstone & Granite

■ ■ ■

"STONE IS A STRONG, TIMELESS MATERIAL which communicates simultaneously the beauty of nature and the handwork of the craftsperson," notes architect **William Leddy** of one of the primary assets of an 1820 stone barn in Pennsylvania that has been adapted to residential use. Partner in the San Francisco firm of Tanner Leddy Maytum Stacy Architects, he says a central element in the design intent was "creating an environment that compresses past, present and future into a living history of place."

To achieve this goal, the one-hundred-and-seventy-year-old stone walls have been kept as the design's centerpiece. After the existing stone was cleaned and new stonework was blended with the old with utmost care, a fascinating mix of other materials, including black Pennsylvania slate and gray Pennsylvania flagstone was added, making the original stonework seem more precious than ever.

OPPOSITE ABOVE: *For the Martin residence in Kennett Square, Pennsylvania, the architects started with an existing barn, with twenty-inch-thick fieldstone walls, situated on ten acres of peacefully rural landscape. Stone wall restoration was by MOBAC, Inc.*

OPPOSITE BELOW: *The inner wood wall and new steel posts and beams carry all structural loads, thus allowing the outer stone wall and existing beams to be exposed frequently. The gray Pennsylvania flagstone, which has been used as exterior paving, continues inside, where it meets bare concrete, a contrast used throughout the house to denote the outer (old) and inner (new) zones.*

ABOVE: *Cantilevered out from its flamed Academy Black granite base, the counter of polished Absolute Black granite makes a play of the light filtering down through a new opening in the second floor. Stone paving and stone veneer throughout are by Banta Tile and Marble Company.*

Photography by Paul Warchol

New additions, such as the new opening which replaces the existing threshing door and opens onto a new deck carved out of the existing porch roof, are frank juxtapositions of the barn's old stone walls with a twentieth-century building vocabulary.

Photography by Paul Warchol

New England Eclectic

Bluestone & Granite

■ ■ ■

THIS NEW ENGLAND SUMMER HOME, nestled on land between pond and marsh, is a comfortable amalgam of American architectural history. Designed by *Mark Simon* and *James C. Childress* of Centerbrook Architects, with interiors finished by *Michael La Rocca*, its curved roofs and dormers recall Nantucket shipwrights' houses. Vertical battening and stickwork which hold large overhanging eaves remind one of the area's Victorian cottages. Arched windows which step up and down within the battening seem both Gothic and modern. Stone walls, appearing as fencing and also used for the main house in monumental bands, are a New England tradition.

Inside, flat stones are like boards and battens at the fireplaces, joining other elements old and new, such as colored plaster, single panelled doors with simple beading, extra-narrow plank oak floors, and baseboards and friezes flush in wood walls. This house is at once old and new, an invention made of traditional elements that invite people to settle inside as comfortably as its exterior settles into the landscape.

RIGHT: *Created from slabs of gray granite, this massively proportioned fireplace centers the house.*

OPPOSITE: *This fireplace has a face of bluestone, which has a purple cast, cut into board-like shapes and configured in a board and batten pattern, as on the home's exterior. Both fireplaces were built by Doyle Construction, of stone supplied by Connecticut Stone Supplies.*

Photography by Jeff Goldberg/ESTO

HONESTLY OKLAHOMA
Prague Red Stone

■ ■ ■

WHEN EXPLAINING HIS APPROACH TO this home, architect ***Rand Elliott*** quotes his most famous fellow statesman, Will Rogers: "If you don't like the Oklahoma weather, just stick around a few minutes and it will change."

Michael and Gerry Reif's residence is a definite response to Oklahoma's ever-changing climate. Its indigenous farmhouse forms, made from a locally quarried sand-stone called Prague Red Stone, green dia-mond-shaped shingles and exposed timber, react to the state's weather with a similar honesty. The north side of the massive stone structure has few openings, pro-viding a psychological sense of shelter as well as physical protection from the cold north winds. The front of the building, on the west side, is composed of an eight-foot-deep porch, offering hospitable shade from the summer's intense westerly sun.

The interior is a simple plan, with the same stone and wood unifying all areas. Plenty of exposed lumber, beadboard planking and knotty pine furniture and a large hand-hewn truss that supports the kitchen's sloping roof further continue Elliott's rugged concept. It is an architecture grown from the land.

ABOVE: *The goal was to build a single-family residence on a wooded site and, through architectural forms expressing the historic flavor of Oklahoma, elevating it to an art form.*

BELOW: *The focus of the living room is the natural stone fireplace flanked by knotty pine paneling and carpenter-scroll cabinetry.*

OPPOSITE: *A red stone wall ties the dining room and living room together into one continuous space.*

**Photography by
Bob Shimer/Hedrich-Blessing**

OPPOSITE: *The stone, native to the state from quarries in Prague, Oklahoma, is layered in a method, often used in the 1930s, which incorporates "jumpers," protruding stones randomly placed.*

ABOVE: *The kitchen, where dinner and family discussions occur around the wood burning stove, has a durable, easily cleanable floor of stained concrete scored into four-foot squares. Durability was also the reason Elliott selected brick for the island facing and oven wall, and granite and ceramic tile for countertops.*

Photography by
Bob Shimer/Hedrich-Blessing

ENGLISH COTTAGE COMFORT

Brick

■ ■ ■

A TREASURE-TROVE OF BRICK WAS already in place at this early-eighteenth-century home on an international dressage horse farm in England, and to designer *Allison A. Holland* it was her main source of inspiration when converting what had been the manager's cottage into a main residence for the owner.

Typical of the brick and flint style used during that period, two types of stonework were used to provide architectural interest to an otherwise flat façade. Each casement window is inset with a border of red brick.

In between the flint, which is a type of sandstone, are horizontal bands of more red brick. Matching chimneys set at either end are reminders that the fireplace was originally the source of heat.

"This architecture so typical of the English horse country had to set the course of action for the interiors," says Holland, who immediately set about removing walls that some previous owner had built to cover up the fireplaces. Today they are once again ablaze, along with Holland's typical assortment of colorful prints and accessories, warming up this part of the English countryside even during the most gray and blustery weather.

ABOVE: *For a change of pace in the family room, Holland decided to coat the brick and flint with plaster while retaining the interesting contour.*

OPPOSITE: *The existing brick and flint facade and brick fireplaces within, already in place at this eighteenth-century cottage in England, served as primary inspiration for the rest of Allison A. Holland's restoration.*

Photography by David Livingston

AUTHENTICALLY TEXAN
Limestone

■ ■ ■

CONSISTENT WITH THE PHILOSOPHY OF the San Antonio firm of Ford, Powell & Carson, this project honors the important connection that can be achieved if a building can be made of material taken from the land on which it sits. To triple the size of a charming, old, one-thousand-square-foot stone cabin, two of the firm's principals, interior designer **John Gutzler** and architect **Chris Carson**, pushed out and up to add a gracious living room, master suite and five guest bedrooms. Yet it is the care with which they treated the original, letting it inspire every aspect no matter how small the addition, that is the main reason for the accolades and awards given this project. The scale of the home as it exists today, and its use of materials from the Texas Hill Country where it is situated, respect the land with all the authencity and joy that ennobled the original cabin when it was built in 1926.

OPPOSITE: *Where regrouting of the existing stone was necessary, the masons used a lighter color to subtly enliven the rooms and amplify the stone's textural quality. The General Contractor was Scantlin Builders.*

ABOVE: *The installation's richly individual broken pattern, highlighted by the local stone's variation in color, celebrates the fact that no two stones are the same.*

Photography by Hickey-Robertson

NATURAL
CONNECTIONS

BALI UTOPIA

Compressed Volcanic Ash, Slate, Marble, Beach Stones & Terra-Cotta Tile

■ ■ ■

ENSCONCED ON THE EAST SIDE OF BALI'S Sayan Valley, this terraced and landscaped property has magnificent views of the peaceful Agung River and surrounding rice fields. Architect **Martin Smyth** and his wife Yajaira welcomed the opportunity to make sure their vacation home, Serkamelati ("Flowering Garden"), would be equally utopian in an equally natural way.

Trained in London with a design philosophy born of the modernist architectural movement, and having worked on massive public projects in Hong Kong, where he is now principal with Steven J. Leach, Jr. + Associates, Smyth found the design process for this paradise a revelation. "It never occurred to me that the first house I would design and build for myself would contain an ethnic quality so far removed from the technological and

ABOVE: *The hard black Batu Karagasan, also known as Batu Besakih, after the famous Balinese temple made extensively of this stone, has been cut by hand and by saw into ninety-three steps ascending from the garden.*

RIGHT: *The pool decks are created from Balimanan, a hard, white and off-white stone from East Java which has been cut by hand. Its low water-absorption makes it quick to dry, and its reflectivity keeps heat absorption low.*

design parameters with which I was familiar," he says.

Throwing himself into an exploration of traditional Balinese construction methods, Smyth developed a symmetrical scheme in which there are no nails or screws, only interlocking timber joints and wooden dowels. Grass fronds densely packed more than one foot thick cover the roof. Four central posts are of shaped and planed coconut palm trunks. And stones, all of them locally sourced in Indonesia, are of such variety and so exotic that even their names evoke a feeling of adventure.

There is white and off-white Balimanan limestone from East Java, the hard, black Batu Karagasan volcanic rock found in Bali, the black slate from North Bali called Batu Singaraja, the yellow-beige Pilah made of compressed volcanic ash, and its dark gray cousin called Paras. These are combined with stones that have yet to be named...Java's local green slate that makes pools look emerald, East

BELOW: *Along the retaining walls made of Pilah, Smyth used the dark gray Paras in hand-cut regular and rectangular facing plates, smoothed and bevel-edged to give the impression of stone block construction.*

Java's loose beach stones of turquoise hue, the highly textured cream-colored marble from Java's Bandung region.... Each of these abundant and natural gifts of the Indonesian geography played a major part in the realization of Smyth's sensitive and soulful vision.

The tile, terra cotta in both color and material content, was produced in wood-fired kilns near the village in which the house is situated. The same sort of tile can be found throughout Bali (hence the local name, Tehul Bali, or Bali tile) and is ordinarily produced in a 290-by-290 millimeter unit (an approximately one-foot-square module including the pointing, or grout) and is fifteen to twenty millimeters thick. This module was used in the "bale," the Indonesian roofed but open-sided room, without wood interlay.

However, a slightly larger module was required to fit between the diagonal timber strips introduced in the living and bedroom floors, so these were made to special order. In order to accommodate their eccentricities of size and shape, the pointing, or grout, was ten to fifteen millimeters.

ABOVE AND OPPOSITE:
Locally made clay tile underscores the feeling of regionalism in various rooms, such as here in the master bedroom, where it is interspersed with strips of wood to create diagonals in harmony with the roof. Balimanan, an off-white hard stone from East Java that is used as exterior paving elsewhere, is used for the bases to the central columns.

Walls of green slate from Java turn a bathroom into an open mini-garden. Inside, beach stones from East Java surround a wood strip floor. Walls are of hand-split marble facing tiles which are rough and highly textured but which have edges smoothed by hand, then fitted without pointing. In the shower enclosure, marble from Bandung is used in slab form, its polished face exposed on walls, its honed face exposed on the floor.

Photography by Carsten Schael

NATURAL REFLECTIONS
Granite & Slate

BELOW: *Juan Montoya established a continuous flow of black slate inside and outside. At one end of the foyer is a painting by contemporary American artist Ellen Nieves and a vase from the Bambara tribe in Africa.*

OPPOSITE: *Highly polished Impala Black granite has been used in monolithic slabs for tables, and in blocks for the fireplace, all designed by Juan Montoya and all gleaming counterpoints to the unpolished black slate floor. Tables are available through Juan Montoya Furniture and Accessories.*

Photography by Jaime Ardiles-Arce

WITH WALLS MADE ALMOST TOTALLY of glass encircling its inner courtyard, this Connecticut home's interior has a refreshing connection with nature which **Juan Montoya** has emphasized by reflecting it. The reflective surface in this case is not mirror but Impala Black granite, so highly polished that it doubles the effect not only of every tree, but of every leaf. The indoor/outdoor quality is extended further by expanses of un-polished black slate for the floor which continue right out to the garden's edge.

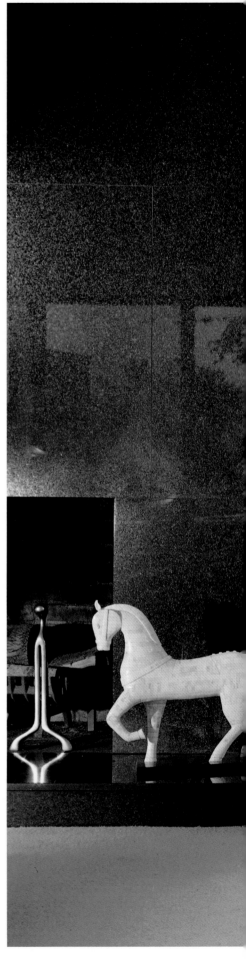

ADOBES DE LA TIERRA
Boulders

■ ■ ■

FOR EIGHTY-FIVE MILLION YEARS, THE awesomely gigantic boulders around which artist **William F. Tull** built this home have commanded this site. Constant reminders of the beauty, character and strength of our natural home, they are the landmarks of Adobes de la Tierra, the community he is designing in Carefree, Arizona.

"The beauty of nature which was here in the first place has a color and majesty unsurpassed," says Tull who, when not building a house here, can usually be found capturing the region's beauty with oil on canvas.

So determined is Tull to not disturb anything in the environment that, whenever he can, he sculpts in adobe around the boulders he finds at each building site. To design living space in and among such

RIGHT: *At the base of the boulder in the master bedroom from which one can look out toward the Sunset Patio, William F. Tull began his plan for this house, drawing, on-site, the rooms he envisioned. "Houses should grow from the earth," he says.*

OPPOSITE: *In harmony with the natural landscape, the house is nestled among gigantic boulders, some of which are eighteen feet high.*

LEFT: *Throughout the house, artist William F. Tull has used a technique he has developed that allows windows to look as if they butt directly against the site's original rock formations, thus allowing them to follow the age-old rocks' majestic contours.*

RIGHT: *A rock formation within which Tull has sculpted a banco and fireplace surrounds and protects the Social Patio off the Great Room.*

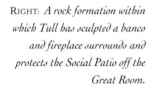

LEFT: *Tull had to remove three or four boulders from the site of the living room, because they did not work with the rest of his design. Yet, in the walls he made niches for even more boulders, as well as three high windows which expose more boulders so that, even from the inside, people can maintain a sense of the site's grand scale.*

Photography by Jennifer Tull

huge rock formations — some are eighteen feet high and that much again in width — Tull and his group of artists/craftsmen have developed many techniques. For windows they use a diamond-blade saw to make grooves in the rocks so that the glass can fit into the groove. To cut the glass to the right shape, a template of the rock is drawn on-site, then set atop the glass, which next is cut to follow the contour of the rock. Before the glass is set into the rock, Tull makes a bed of waterproof silicone mixture, which he places in the grooves, then camouflages with a mixture of mud or rock taken from the site, mixed with more of the silicone material. In the end, it cannot be seen, and the glass seems to connect directly to the boulders.

"You don't want to tamper with some of the most beautiful things on earth and any altering you have to do should be invisible," says Tull.

AQUATIC DELIGHT
Ceramic Tile, Glass & Quarry Tile

■ ■ ■

SOMETIMES A SENSE OF DEPARTURE IS more important than complete continuity. After *Alfredo De Vido Associates* had completed the renovation of their existing early-twentieth-century house, Thomas and Eileen McConomy asked the architects to add an enclosed swimming pool as well. The architects harmonized this necessarily large structure with the house by using a similar brick and slate outside — but within, the poolhouse has a character all its own.

A pattern of skylights brings in natural light in the daytime, while night lighting enhances the space still further. Both day and night, the sparkle of brightly colored glass tile evokes the spirit of joyful exercise.

For the walls, the architects created a decorative pattern of Hastings glass tile and American Olean ceramic tiles, combined with Drywit — all moisture-impervious materials so important in humid conditions. The floor is American Olean quarry tile with an anti-slip finish.

Photography by Lockwood Hoehl

ECHOES
FROM
CENTURIES
PAST

ABOVE: *The use of cantera stone inside and outside begins at the top of the fountain and continues its descent all the way to the front door. The façade is also of cantera stone.*

OPPOSITE: *Welcoming guests to the foyer is a Venetian mosaic of glass and marble on a background of Absolute Black granite surrounded by crackled walls with trompe l'oeil paneling. The columns have faux marble pedestals with Absolute Black granite pilasters. The top of the iron and acrylic console also is faux marble. All are spectacularly illuminated by a Murano glass chandelier. Marble and granite throughout were supplied by Fachadas y Monumentos. Venetian mosaics throughout were created by Byzantium.*

INSPIRED BY THE CLASSICS
Cantera Stone, Granite & Marble

■ ■ ■

STONE IS THE SOUL OF THIS RESIDENCE by *Samuel Sandler* of Mexico. It is used everywhere and in many forms — in keeping with this architect's philosophy that ornamentation helps people relate to the spaces in which they live. "And if they can feel good where they live, if they feel content, they'll be happier and healthier," he says.

Sandler was asked to design the architecture and interiors for clients whose children are grown and who were most interested now in having a place where they

RIGHT AND BELOW:
Grayish-white cantera stone, a color of this Mexican sandstone that is rare, is used throughout the interiors, along with a profusion of trompe l'oeil to provide a feeling of further openness under the floating vaulted ceilings. All the stone was hand-carved by Mexican artisans, and in the television room the blocks and molding had to be sculpted to fit the circular shape of the space. Main points around the axis, dramatically defined by polished granites and cantera stone, are entry hall, master bedroom hallway, and television cabinet. The cantera stone throughout was supplied by Arte En Cantera.

OPPOSITE: *In the powder room, the pure white of the Pentelicon marble countertop is echoed in the tassels as well as moldings carved by hand from plaster.*

could entertain. Of interest is that they had wanted a contemporary house, whereas Sandler's firm is geared toward classical design. A compromise was reached: The architecture would be classically inspired and the furnishings would include some contemporary pieces. The result is an eclectic mix that is elegant but not intimidating — definitely an atmosphere in which both clients and their guests can feel very good indeed.

ABOVE AND OPPOSITE: *In the living room, the Italian limestone floor with marble border, cathedral ceiling defined by the hand-carved plaster, accessories such as the hand-carved marble frieze and muted colors help achieve the sense of elegance the client desired for formal entertaining. Limestone supplied by Fachadas y Monumentos. The marble border was through Byzantium.*

Photography by Julio Tovar

RENAISSANCE REDISCOVERED
Limestone, Travertine & Marble Mosaic

∎ ∎ ∎

HAVING BEEN RAISED AND EDUCATED IN Europe and Asia, where she has worked as a designer for much of her life, **Renée Kubiak** brings an international complexity to her interiors that always goes beyond the obvious, even in the smallest spaces. In this four-foot-by-six-foot bathroom in Orange County, California, she worked with a Parisian to carve the sink and a Spaniard to create a traditional Italian mosaic — making Kubiak's interpretation of "the Renaissance rediscovered" quite literally a global affair.

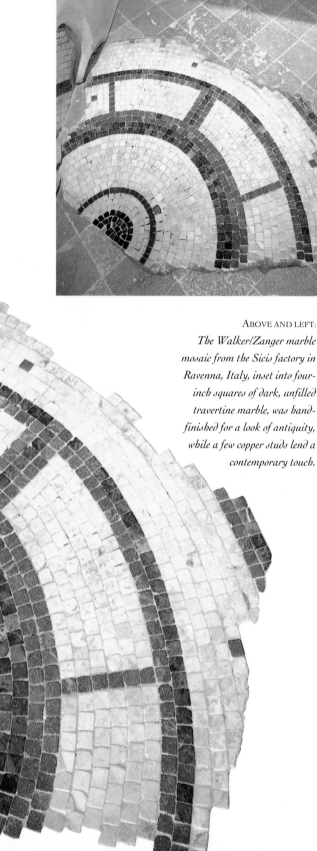

ABOVE AND LEFT: *The Walker/Zanger marble mosaic from the Sicis factory in Ravenna, Italy, inset into four-inch squares of dark, unfilled travertine marble, was hand-finished for a look of antiquity, while a few copper studs lend a contemporary touch.*

OPPOSITE: *The sink, which cantilevers out so that it appears to float, was carved from a three-and-a-half-inch-thick, seven-hundred-pound slab of French Beaumaniere limestone. The bowl was cut by hand and is only one-and-a-half inches deep, creating the illusion that it has been naturally eroded by water. It was designed by Kubiak, in association with Stacey Dukes and Charles Gray. The limestone was supplied and fabricated by Impression.*

Photography by Bud Lammers

ROMAN BATHS RECOLLECTED
Limestone, Stone Mosaic & Travertine

■ ■ ■

THE ARCHITECTURE OF ***Oscar Tusquets Blanca*** seems as much sculpture as design, and perhaps nowhere is this more evident than in the main bathroom he created for a single-family dwelling in Madrid. The exterior itself is reminiscent of Pompeii, its orchard of Roman gardens. So perhaps it is not surprising to find inside an area that evokes thoughts of the Baths of Caracalla.

In another bathroom that he designed for his own home, the Villa Andrea in Barcelona, ancient Rome is again recalled, this time in a first-century Roman mosaic.

RIGHT: *By placing a first-century Roman mosaic in the center of the floor, Oscar Tusquets Blanca turned this bathroom at his Villa Andrea into a work of art. The limestone is from Palancar.*

Photography by Rafael Vargas

ABOVE AND LEFT: *Bas-relief designed by Barcelona architect Oscar Tusquets Blanca and carefully sculpted in situ of travertine marble by Juan Bordes, this fireplace, tub and vanity function for bathing and dressing... but more important is that they are works of art to enrich daily life.*

Photography by Jordi Sarra

ARABIAN NIGHTS
Ceramic Mosaic Tile

■ ■ ■

ARCHITECTURE BUFFS, WHO SEEM to include the majority of the residents of Berkeley, California, cherish elements of fantasy and romance in older buildings, and when Xanadu was renovated, such romantic fantasies became reality. One of the least remarkable 1940s homes in the area, it has been transformed by Ace Architects' *David Weingarten*

ABOVE, LEFT
AND OPPOSITE ABOVE:
*The Historic Color Guide,
a 1930s handbook, provided the
basic palette for the house, a set
of colors known as "Persian
Miniature." At the entry, they
are placed in a spiral pattern of
McIntyre Tile's one-inch gold
tiles, used sparingly and set into
American Olean's "Ali Baba"
blue and cerulean blue tiles.*

Photography by Alan Weintraub

and ***Lucia Howard*** into an assemblage of domes, arches, columns, and tilework that reverberates with childhood memories of Sinbad and the Arabian Nights.

RIGHT: *At the fireplace, the small gold and blue tiles further allude to Islamic tile patterning.*

THIS PAGE:
Once the old walls were revealed, the architect carefully built them up again with layered planes of frescoed plaster and bleached ash, creating the impression that layers of history had been peeled away.

NEW YORK ARCHAEOLOGY
Brick, Marble, Ceramic & Glass Mosaic Tile

THE CONTEXT OF THIS WEST VILLAGE residence enthralled architect *Vishva Priya* from the start. Located in an apartment house in what was once a meatpacking plant built on top of early-nineteenth-century houses of New York City's Old Amsterdam area, it had so many unexpected treasures beneath a developer's recent addition of gypsum board walls, that he used what he found to direct the character of the total renovation.

In the process of combining two apartments into one, he discovered underneath the existing plaster such a beautiful collage of random, uncoursed brick and stone that demolition became archaeology. The walls were carefully cleaned by hand with muric acid, brushed with wire bristles and washed with soap and water. Comments Priya, "We felt we were dealing with something very precious."

In the bathrooms, Priya did an about-face, using new tile to turn them into archaeological jewels of a different order.

ABOVE: *Various ceramic and glass mosaic tiles create an abstract geometry of color and texture in these minimalist bathrooms by Vishva Priya. Green and gray glass mosaic tile is from Vetricolor. Beige marble tile is Bottocino Classical, imported from Italy by Marble Techniques.*

Photography by Michael Moran

MYTHOLOGY IN AFRICA
Marble, Stone & Marble Mosaic

■ ■ ■

THEY WERE ASKED FOR A DREAM —to create an other-worldly home-away-from-home in Bophuthatswana, southern Africa — and interior designer ***Trisha Wilson*** and architects ***Wimberly Allison Tong & Goo*** turned the assignment into a reality more fantastic than a dream could ever be. The Palace of the Lost City is a park and hotel that includes elaborate interpretations of the architects' development of a legend about an ancient city they have "uncovered," and much of the living mythology has been created from marble and stone.

LEFT: *In order to achieve the mix of colors, the designers specified marbles from around the world, including pale yellow Crema Valencia marble with blood-red veins from Spain, Rosso Verona marble from Italy, African Black Quagga and African white marbles from Namibia.*

Photography by Peter Vitale

Below: To achieve the lively scene of the procession of animals on the entry rotunda floor, more than seventy types of marble were used. The elephant alone is made of approximately three-hundred-fifty-thousand pieces of more than fifteen varieties of Italian marble for the different colors of gray.

Above: The stone mosaic inset into the wall at the base of the grand stairway is made of twenty-four semiprecious stones — including lapis lazuli, tiger's eye, topaz, pink quartz and malachite — all from the African continent. The floor is Crema Valencia marble. Marble throughout was supplied and installed by Taung Marble and Marblelime, Johannesburg. Mosaics throughout are by Sfreddo & Delgalo, Pretoria.

LONDON & ST. TROPEZ
Marble, Stone Mosaics & Quarry Stone

■ ■ ■

STONE USED IN DRAMATICALLY DIFFERENT ways is a hallmark of interiors by *Tessa Kennedy*. In the renovation of one home in Cottesmore Gardens in London, the range of types and sizes of stone creates great variety from room to elegant room, and even a swimming pool that looks like a Roman bath. In the design of a speculative apartment in Lowndes Court, the grandeur of her octagonal marble entry hall makes one think of royalty. Yet, in her transformation of a farmhouse in St. Tropez into a villa for the late film producer Sam Spiegel, her choice of stone from a local quarry exudes a sense of regional authenticity.

ABOVE: *In transforming a St. Tropez farmhouse into a villa for the late film producer Sam Spiegel, Tessa Kennedy emphasized the home's original character by using a local quarry stone for all floors and stairs.*

LEFT: *To achieve a spacious look for the entry to a speculative apartment in Lowndes Court, London, Tessa Kennedy turned what had been a hall into an octagon and selected a pale color scheme, even underfoot. The marble was supplied by Reed Harris.*

OPPOSITE: *A marble-inlaid floor at the entry of Cottesmore Gardens is the first of Tessa Kennedy's variations on the elegant theme.*

For the Cottesmore Gardens
swimming pool, Tessa Kennedy
completely gutted the basement to
make room for walls of marble slab
paneling relieved with stone mosaic
designs copied from ancient
Pompeii. All stonework at
Cottesmore Gardens is by Walter
Jenkins & Co. Ltd. The mosaics
are from Domus Tiles.

A SENSE
OF TRADITION

TUSCAN-STYLE VILLA
Sandstone, Limestone, Flagstone, Travertine & Slate

■ ■ ■

ONE WAY TO BRING CHARACTER TO a nondescript stucco structure is to endow it with more glamorous materials. When asked to completely redesign such a home in Los Angeles, **Mark Warwick** and **Kim Hoffman**, principals of The System Design in Beverly Hills, decided to incorporate an abundance of various stones in the style of a traditional Tuscan villa. They used almost every kind known to those classical Italians, and today one would never know it was the same ugly duckling.

ABOVE: *All exterior stucco walls were capped with cast stone manufactured by Adriatic Cast Stone and painted to create an aged patina. Entry steps are Rosa Adoquin tiles from Mexico through Bourget Bros. Coast Flagstone.*

RIGHT: *A stone fountain incorporates plinths, columns, a bowl and coping in cast stone and plaster. Fountain imported from Italy through Sculpture Design Imports.*

LEFT: *The entry's floor of honed rather than polished travertine further creates the feeling of age. Travertine throughout from Globe Marble & Tile.*

BELOW: *The pool is edged with black slate to create the effect of a reflecting pool. Six antique stone orchid planters, imported from Mexico by the designers, feed the English ivy on the wall beyond. Decking is Arizona Flagstone from Bourget Bros. Slate is through Globe Marble & Tile.*

Photography by Patrick House

RIGHT: *A custom-designed cast-limestone fireplace keeps company with a Rojo marble tabletop and a custom sofa table with inset limestone. Pre-cast stone through Adriatic Cast Stone. Tabletops through Globe Marble & Tile.*

BELOW: *The dining tabletop is Porto limestone through Globe Marble & Tile. All casings, crown moldings and baseboards have been faux-finished to convey the feeling of stone.*

RIGHT: *In the master bathroom, travertine is used for flooring as well as the tub surround and shower enclosure.*

OPPOSITE RIGHT AND ABOVE: *While the guest bathroom was limited in terms of space, two kinds of limestone for flooring, wainscoting, and the sink create a sense of grandeur. The limestone, supplied by Rhomboid Sax Bath & Tile, was cut by hand to replicate various widths and lengths more commonly found in Europe.*

CALIFORNIA HACIENDA
Flagstone, Granite, Limestone, Slate & Syndecrete

■ ■ ■

WHEN THEIR CLIENTS WANTED THEIR HOME in the Southern California desert to feel like a Mexican hacienda, interior designer **Barry Brukoff** and architect **Richard Stowers** decided that stone — lots of it and of various kinds — would be integral to their concept.

"The intention was to create a feeling of timelessness and warmth through a rich palette of natural elements," explains Brukoff. "Stone combined with copper, wood and multi-colored planes of stucco provides a sense of traditional Mexico, complementing the home's contemporary structure."

Of major importance is the home's floor, inside and out of slate from India that was ordered in a custom size of twenty-inch squares. Anticipating the many months' lead time required, Brukoff ordered the material far enough in advance of installation that he was able to specifiy the large size without incurring additional cost.

RIGHT: *The tones of the slate floor influenced Brukoff's color choice of Syndecrete, a stone aggregate concrete which gave him an economic alternative to granite for the custom bar counter.*

OPPOSITE: *Inside and out, slate from India in a large custom size of twenty-inch squares provides a unifying element underfoot. The slate was imported by Eurocal Slate Centers. The floors were installed by Bogart Masonry. The general contractor was Stoker Construction Inc.*

Photography by Paul Stowers

BELOW: *French limestone from Natural Stone Ltd. for the kitchen counters provides a durable surface while maintaining the organic palette.*

OPPOSITE: *The decision to continue the slate to the exterior patio creates a sense of space and continuity, providing a feeling of expansiveness to both interior and exterior.*

ABOVE: *In the dining area, Brukoff inset pieces of the same size slate that was used for the floor into the top of the buffet cabinet which divides living and dining areas.*

LEFT: *In the master bedroom, the warmth and texture of a tribal kilim softens the cool quality of the slate.*

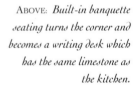

ABOVE: *Built-in banquette seating turns the corner and becomes a writing desk which has the same limestone as the kitchen.*

RIGHT: *Above and below the fireplace, Arizona flagstone in an intricate running bond pattern using two different sizes of stone creates an unusual design of natural hues. The flagstone, which is a kind of sandstone, as well as the granite on either side of the firebox is from Natural Stone Ltd.*

THEME & VARIATIONS
Used Terra Cotta & New Hand-Painted Ceramic Tile

■ ■ ■

TILE BECOMES SYNONYMOUS WITH romance in the lyrical work of **Samuel Botero**. To him, the patina of old tile can inspire a symphony of complementary values, while he rhapsodizes over the way the vast number of glazes, shapes and sizes available in contemporary tile makes the variations possible on any given theme seem endless.

For one home's entry and solarium, he dresses a collection of Victorian furniture in floral chintz and sets it dancing over a floor of antique terra cotta and wood from an old château in France. In

RIGHT AND OPPOSITE:
For the installation of this antique floor from France, it was necessary to first sand and stain the wood, then finish it with a coating of polyurethane. Next, the original grout, which had not been removed from the tile, had to be made as smooth as possible so it could fit within the wood grid. Finally, a sealer was applied to the tile.

Photography by Dick Busher

the bathroom of another home, he composes an old-fashioned duet, combining uneven handmade tile, which has been covered with delicate hand-painted wisteria and glazed, with large, unglazed squares of terra cotta. Yet down the hall in another bathroom, to pick up the rich reds in a kilim rug, he features a smooth, precisely cut machine-made tile.

"I love them all," says this New York designer.

LEFT: *The rich reds in a patchwork carpet made from kilim remnants led to the selection of tile in this bathroom. The smooth, precise cut of this machine-made tile allowed for minimal grout. The predominant tile is from Hastings Tile Company. The small unglazed mosaic tile in the shower is from American Olean.*

OPPOSITE AND LEFT:
A wide grout was necessary to accommodate the inconsistencies of this handmade tile, but designer Samuel Botero says he did not mind: "A wider grout contributed to the old-fashioned look I was after." Both the hand-painted tile and the terra-cotta floor tile are from Hastings Tile Company.

Photography by Phillip H. Ennis

EUROPEAN INSPIRATION
Hand-Painted Ceramic Tile

■ ■ ■

WHETHER INSPIRED BY AN EIGHTEENTH-century Spanish kitchen or a seventeenth-century Dutch painting, interior designer **German C. Sonntag** makes visual music through a myriad of design themes that reach their height in his use of hand-painted ceramic tile.

He comments, "The innumerable combinations that can be obtained on a material so durable make it an endlessly exciting receptacle for a room's major color, pattern and figurative elements."

ABOVE: *An eighteenth-century kitchen displayed at the Museo de la Ceramica in Valencia, Spain, inspired this imaginative design for the Russell and Norma Hanlin residence in Pasadena, California. German C. Sonntag selected red granite from Roma Marble for the island top, not only because of its color and grain, but also for its durability. Countertops and splash were decorated with painted tile from Walker/Zanger which has been set in a random pattern that complements the colorful interior. The hood is covered with tile with a hand-painted basket motif by Jeannee's Custom Tile.*

OPPOSITE BELOW AND BELOW: To turn a utility kitchen into a contemporary gathering area for a family, German C. Sonntag selected a wood floor with a decorative tile inset border to underscore the feeling of warm intimacy that he wanted to achieve. The color of the motif in the floor was then used in the grout to obtain an interesting pattern in the countertop and splash of more randomly placed hand-painted tiles by Melinda Strauss. The windowsill and front of the counter are finished with a decorative tile that has a linear motif. Over the commercial range is a reproduction on tile of a seventeenth-century painting by Jan Van Huysum. The tile was supplied by Latco Products and installed by Ralph McIntosh.

Photography by Leland Y. Lee

HONOLULU INFORMALITY
Salvaged Brick

■ ■ ■

IN HAWAII, ONCE MISSIONARY SETTLERS realized that the climate does not require building materials of such heft, brick ceased being used in homes. Yet one can still find antique brick from some of those early-1800s buildings which have since been razed, and *Allison A. Holland* has capitalized on its individuality.

For the living room of this newly built family residence in Honolulu, which she and architect *George V. Hogan* scaled for intimate conversations, music and reading,

she selected salvaged brick for the mantel to underscore the sense of informality and warmth. "The organic, earthy colors in old brick provide psychological warmth while, in fact, lowering the temperature," says Holland. "Brick is just as good for keeping heat out in warm climates as it is for keeping it in where climates are cold."

On the lanai, the brick floor, which is laid in a crosshatch pattern, wards off the sun's heat as well as provides an im-

mensely practical surface which can be hosed down.

For the hospitable atmosphere created by the use of salvaged brick and other complementary materials, as well as for this home's suitable and charming sense of scale, this project won a first place award in a competition sponsored by the American Society of Interior Designers.

OPPOSITE AND RIGHT:
The texture and color of salvaged brick make it a fascinating material to use in conjunction with wood, printed fabrics and wicker for a look of relaxed and intimate comfort.

Photography by David Livingston

SPANISH ELOQUENCE

Fieldstone, Granite & Marble

BELOW: *Spanish and Italian marbles welcome guests to the eighteenth-century palace of the Viscount of Miralcazar.*

Photography by
Francisco de la Fuente

■ ■ ■

THAT *Joaquin Alvarez Montes* EXCELS IN theater and film design as well as in architecture and interiors is evident as soon as one enters the eighteenth-century palace in Spain inherited by the Viscount of Miralcazar. Beyond the purely technical aspects of having to completely gut and modernize the wood structure, which was in great disrepair, he brought it to life by emphasizing its palatial aspects in some areas, while, in others, interpreting his client's character as if he were creating another set.

For public areas such as the vestibule, where guests are first welcomed to the interior, the viscount wanted to maintain the traditional style of the palace, and Alvarez Montes selected only the noblest materials. For the walls, a mixture of cement and marble dust which has been highly polished by hand is joined by reams of cut velvet. On the floor, Spanish and Italian marbles grandly underscore this elegant prologue, their color and reflective quality further enhanced by polish applied in varying amounts so as to modulate the visual perception of depth.

In the client's private quarters a completely different approach reflects the young viscount's request for a more contemporary look. Explains the designer, "The bath, for example, reflects my client's preference for a masculine, warm feeling in contrast to the grandeur elsewhere."

From the family farm, Alvarez Montes selected fieldstone for the bathroom fireplace, and in Granada he found a red-brown granite for the countertop and

bath. Combined with a herringbone wall-covering from Gene McDonald in New York and a carpet that, like the carpet on the vestibule's stairway, was made by hand at Real Fabrica de Tapices in Madrid, the look is unquestionably male.

RIGHT: *In the foyer, which is the centerpoint of ten reception rooms, the noble marbles continue their reflective display, enhanced by varying degrees of polish.*

LEFT: *The private bath of the viscount, made from a hallway and storage room, incorporates regular and tinted mirrors to visually extend the space and complement the masculine strength of the fieldstone and granite. Marble and stonework throughout are by Angel Dilla Moline. The general contractor was Andres Minero.*

EUROPEAN SHAKER DUET
Mosaic, Hand-Painted & Custom-Colored Ceramic Tile

▪▪▪

IT WAS TO BE A SPECIAL PLACE FOR GUESTS of Howard and Roberta Ahmanson, and with every single tile in its kitchen and bathrooms being custom designed, its expression of hospitality is exceptional. *Joseph Alcasar Terrell*, project designer with design control of all elements of construction and finish, has combined strong hints of pared down Shaker-style design from the northeastern United States while supporting a feeling of Europe from centuries past. "The entire house has a sense of architectural history with modern organizational elements," he says. It works and it works well.

ABOVE AND LEFT: *A custom designed ceramic mosaic, the center of attention in the master bath, is echoed by the custom mosaic sinks. Also custom are the ceramic vine borders along the splash and tub/shower. Tile provided by Concept Studios.*

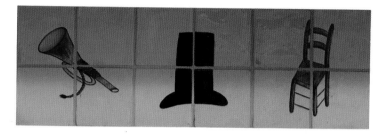

ABOVE: *The reserved use of custom hand-painted tiles and their folk-art design combine with the Shaker-style cabinetry to instill a feeling of simplicity. Tile from Concept Studios.*

RIGHT: *In the guest bath, pure white tiles, liners and border convey a sense of absolute calm. Tile from Concept Studios.*

Photography by Grey Crawford

ARTSCAPES

SHOWCASE FOR ART
Limestone

■ ■ ■

A COLLABORATION BETWEEN INTERIOR designer **Sheri Schlesinger**, architect **Jaime Gesundheit** and their clients, this home was to be filled with light, with barriers between inside and outside minimized. For both, Beaumaniere limestone throughout — for paving, stairs, counters, fountain, banquette seating, fireplace surrounds and even the pool coping — was key. Says Schlesinger: "The off-white hue of this stone provided the color palette for the entire space, and its weighty character provided the strength, both physical and psychological."

ABOVE: *A sense of light begins at the exterior by introducing the Beaumaniere limestone and highlighting it with a metal sculpture, "Enigma" by Paul Mount. Limestone flooring throughout has been supplied by Walker/Zanger.*

LEFT: *Stone coping around the pool is echoed in a stone and steel sculpture by Woods Davey.*

OPPOSITE: *Grounded on a sea of limestone and surrounded by light and white, the home provides an ideal showcase of art, including (clockwise from left): "Chieftan" (brass) by John Huggins; "Shell" (wood) by Tom Carr; "Guifa e la Berreta Rossa" (1989 aquatint) by Frank Stella; and "Posed Piety" (oil on canvas) by Jonathan Santofler.*

LEFT: *In the living room, the light muted tones of the limestone and all other furnishings continue to provide a foil for art. Over the fireplace is "Flying Man with Briefcase" (painted Gatorfoam) by Jonathan Borofsky. On the far wall behind the piano is "Black Spinette Under Arch" (cellvinyl acrylic nova gel and dry pigment) by Ron Davis.*

TOP: *Renaissance Marble fabricated all horizontal surfaces of limestone, here serving as neutral background for the brilliantly colored "Haida Gloves" by Ed Pashke. At left is "Weeping Woman" (black marble) by Vinni Poo-Nampol.*

ABOVE: *The cast bronze frieze above the planter is by Robert Graham.*

ABOVE: *Like multicolored jewelry against a sea of black granite are Peter Shire's "Scorpion Orange Ring" teapot and Franco Assetto's acrylic-on-plastic candies. Italian granite counters were fabricated by Renaissance Marble.*

OPPOSITE: *In the powder room, limestone from Walker/Zanger continues up the wall, and even the bowl has been carved from a block of limestone. Fabrication: Renaissance Marble. Art: "Redstart" by Dan Christiansen.*

Photography by Mary E. Nichols

LIGHT PLAY
Limestone, Granite & Marble

■ ■ ■

LIKE A CALIFORNIA PLEIN-AIR PAINTING, this home is a study of its region's fascinating light. **Reginald Adams** seized the opportunity when presented with his interior canvas — a structure designed with one space opening up to the next through lyrically arched openings, and with various ceiling heights and stepped-down floor areas rather than four walls and a door defining the rooms. It was evident to him that he should select neutral light colors, to augment rather than interrupt the feeling of spacious continuity, and concentrate on the reflective character of paint, stone and faux stone to enhance the aura of unencumbered lightness.

LEFT: *Making a still life of pale hues are French limestone floors with polished Absolute Black granite accents from JB Marble, a collection of stone artifacts from Formations, a faux stone urn and pedestal from Michael Taylor Designs, and columns with a special painted-stone finish by Russell White.*

LEFT: *The entry's stone materials flow into the living room with its French limestone flooring, crown molding finished by Russell White to look like stone, Gina B.'s iron side table with beveled marble top, and a custom coffee table with Absolute Black granite top inlaid with Rojo Alicante marble and Perlato marble. Other faux stone elements are Thomas W. Morgan's floor lamp, Formations' Capital table and the table in the bay window painted by Russell White to look like stone.*

RIGHT: *Carved table bases of volcanic stone hold a polished Absolute Black granite top. The centerpiece is an ancient Japanese stone on a carved presentation base, both from J. F. Chen Antiques. The Absolute Black granite pedestal is from JB Marble. The reproduction Roman stone urn is from Dennis & Leen.*

Photography by Mary E. Nichols

Ancient stone and pottery fragments from J. F. Chen atop a faux stone table finished by Russell White.

VILLA SANNARI, HELSINKI
Hand-Painted Glass

■ ■ ■

ABOVE: *The artwork in the yard is incorporated into the square columns which are sheathed with Liukko-Sundström's one-of-a-kind ceramic tiles, similar to the ones which for many years have displayed her art on the wooden fence in the background. The vertical areas in the facade consist of hand-decorated ceramic tiles of standard size, while the upper section consists of decorator tiles of various sizes.*

The tiles are delivered to order in single pieces and are installed on-site with a special mortar. The stationary green flower boxes are covered with standard-size tiles, but the green glazing and decorating were done to order as one-of-a-kind artwork. "Opening Flower," the motif incorporated in the facade, is repeated on the glass detail of the door leading to the foyer.

WHEN FINNISH CERAMIC ARTIST *Heljä Liukko-Sundström* created Villa Sannari for her daughter's family in an old section of Helsinki with the help of architect *Leena Yli-Lonttinen*, a new application for the ceramic art produced by Arabia was launched. The main construction materials of the residence are stucco of two different colors and Siporex, a lightweight and sturdy hardened concrete with the strength of stone. Throughout Yli-Lonttinen's contemporary design, Liukko-Sundström's one-of-a-kind tiles have been used. They emphasize the entry door opening, enliven the portal and balustrades, and individualize every room.

With works in the collections of the Victoria & Albert Museum in London, and the Boyman Van Beuningen Museum in Rotterdam, Liukko-Sundström is known for her art's joyful, life-enhancing quality, evident here in the feeling of sunshine her work brings to this house regardless of the weather.

The tiles designed by Liukko-Sundström either belong to Arabia's standard production or are one-of-a-kind works of art, but the ideas can be applied to the enhancement of any home. As at her daughter's home, such use of tile can enliven the landscape and bring delight to its owner and also to everyone else who sees it.

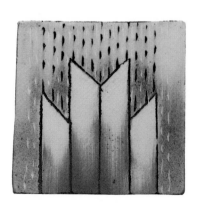

RIGHT: *Decorative ceramic tiles on risers add color and visual interest to the sleek white expanse of stairs.*

BELOW: *On the columns in the foyer, a row of one-of-a-kind "Bud" ceramic tiles is embedded into the Siporex arch-ingots. Four larger tiles have been added in the middle of the tile floor.*

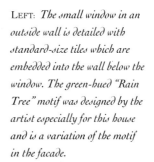

LEFT: *The small window in an outside wall is detailed with standard-size tiles which are embedded into the wall below the window. The green-hued "Rain Tree" motif was designed by the artist especially for this house and is a variation of the motif in the facade.*

BELOW: *"Blue Joy," a painted-glass piece by the artist, is embedded in the terrace wall.*

LEFT: *The windowsill of the nursery is covered with the artist's ceramic ode to teddy bears.*

OPPOSITE: *A staircase balustrade made of Siporex is detailed with openings in which light fixtures are hidden behind Liukko-Sundström's glass art.*

Photography by Timo Kauppila/Indav

LEFT: *A false decorative half-column made of ceramic tile, natural carved stone and terra cotta finished in a custom-colored limestone paint is one of a pair that flanks the entry to the master bedroom cottage. Imported and domestic tile throughout is from Sunny McLean.*

Photography by
Dan Forer

VILLA MALAGA, FLORIDA
Ceramic, Terra Cotta & Mosaic Tile & Cantera Stone

■ ■ ■

THE FEELING OF BEING AT ONE WITH THE land envelops you when entering Villa Malaga, the home of tile merchant Sunny McLean and interior designer ***Dennis Jenkins*** in Coconut Grove, Florida. The steamy subtropical climate is so undeniably reflected through their abundant and various uses of tile that it is sometimes difficult to distinguish their exuberant walls, floors and patios from the exotic flora and fauna surrounding them.

OPPOSITE ABOVE: *In a patio, both floor and bench are finished with black and white ceramic tile and natural cantera stone. The bronze plaque is surrounded by shards of copper-glazed ceramic tile.*

OPPOSITE LEFT: *The mellow but exotic subtropical feeling is expressed through a bronze plaque urging, "Sotto Voce," surrounded by copper-glazed tile shards.*

ABOVE: *A serpentine wall of ceramic tile shards winds through the south garden.*

LEFT: *Copper tile shards cut into the face of a garden wall.*

ABOVE: *Terra cotta, with its reverse side exposed, is combined with French blue ceramic tile around a hand-carved teaching tablet from the Euphrates and inset into squares of two-inch-square terra-cotta tiles.*

RIGHT: *A portion of the "Story of Life," mosaic, twenty-one feet long and nine feet high, is made of individual tile plaques and shards of multicolored ceramic and terra-cotta tile.*

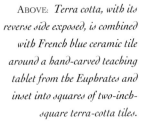

OPPOSITE RIGHT: *The custom-designed kitchen counter of black and white checkerboard tile is framed by a purple heartwood bar rail and free-form design of broken ceramic and terra-cotta tiles.*

RIGHT *Detail of the library's ceiling mosaic.*

Photography by Lanny Provo and Dan Forer (right)

LEFT: *A wall glistens with copper-glazed tile.*

EARTH SPIRITS IN SANTA FE
Reused Ceramic Tile

■ ■ ■

PERCEIVING EVERY COMMISSION AS AN opportunity to express the spirituality of man and nature in an ecologically suitable way, **James F. Jereb** uses only recycled materials, many of them bits of colorful ceramic tile, to sculpt his vision. As did the ancients in his world of Santa Fe, New Mexico, and in those foreign lands he ceaselessly explores, he looks at every wall as an art form, at every crack as a chance to create some new delight.

Jereb's six-foot-long turtle of tile, mirror, adobe and cement gazes out from a twenty-five-foot-high wall.

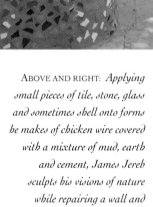

ABOVE AND RIGHT: *Applying small pieces of tile, stone, glass and sometimes shell onto forms he makes of chicken wire covered with a mixture of mud, earth and cement, James Jereb sculpts his visions of nature while repairing a wall and creating a fountain and patio.*

ABOVE RIGHT: *Inhabiting a patio is Jereb's eight-foot-long lizard made of tile, mirror, adobe, cement —and lots of spirit.*

Photography by Lynn Lown (opposite and above right) and Robert Reck

N EW

H ORIZONS

BOLD & MONOLITHIC

Corian®

■ ■ ■

WHEN DESIGNER *Warren H. Snodgrass* planned the kitchen and family room for his new home in Ross, California, his goal was to create a clean, contemporary space. However, since his family includes two teenage boys, the space needed to be easy to maintain and virtually indestructible. The answer — DuPont Corian®, a high-performance acrylic material that can be worked as easily as fine hardwood yet has the translucent quality of ceramics.

"I selected Corian® not only because it is impervious to burning or staining like tile and stone. I also could use various techniques to form, bend and shape it without a single visible seam," says Snodgrass, whose design philosophy is based on clean, unadorned shapes, free of superfluous decoration.

The material seems ideally suited for this designer's bold, monolithic shapes.

The countertops and work surfaces are strong, uninterrupted forms that, particularly in the case of some free-standing pieces of furniture, appear to be sculpted from a solid block of Corian®.

Notes Snodgrass: "The designs, a blend of sharp edges and soft curving surfaces, would be difficult to achieve with any other material."

For his kitchen and family room, Warren H. Snodgrass' selection of pure white DuPont Corian® enabled him to create seamless as well as virtually indestructible countertops and work surfaces, as well as several pieces of furniture.

Photography by Chas McGrath

BACKDROP FOR ART & LIFE
Norman Face Brick

■ ■ ■

FOR CENTURIES USED FOR ITS DECORATIVE, insulating and durable surface qualities, brick was used as a full structural partner in this Case Study House designed by **Donald Hensman** and the late **Conrad Buff III** for the constructive and innovative program of *Arts & Architecture* magazine in the 1960s. By using Norman face brick piers as the major elements throughout the house for both vertical and lateral loading, the architects demonstrated the value of brick as a structural material that requires no finish for either interior or exterior use.

Face brick, the variety which is more refined and dense than common brick and incorporated here both inside and outside for all floor and wall surfaces, not only achieves a harmonious unity through-out, but also creates an exceedingly mainte-nance-free environment.

The design utilizes the entire site, with two rectangular forms linked by glass-enclosed galleries forming a central pool area open to the clement weather of Southern California. In the interior, completed by Robert D'Amico in association with Buff & Hensman, the brick sets the tone for a *concept emphasizing natural materials and provides a warm but not competitive backdrop for works of art. The brick was supplied by Pacific Clay Products.*

Photography by Julius Shulman

A Change of Pace

Ceramic Tile

■ ■ ■

A CONCENTRATION ON LIGHT, VOLUME AND color, and the elimination of detail mark the work of architect **J. Frank Fitzgibbons**. "In this era of diminishing craftsmanship and budgets, I feel it's my responsibility to find ways other than elaborate ornamentation to make environments nurturing," he says — and find them he does, especially in his use of tile in this Southern California home. Throughout he has used the same black ceramic tile to contribute to a sense of harmonious unity, yet he has varied the tile's finish, size, and even the color of the grout to create diversity within that unity.

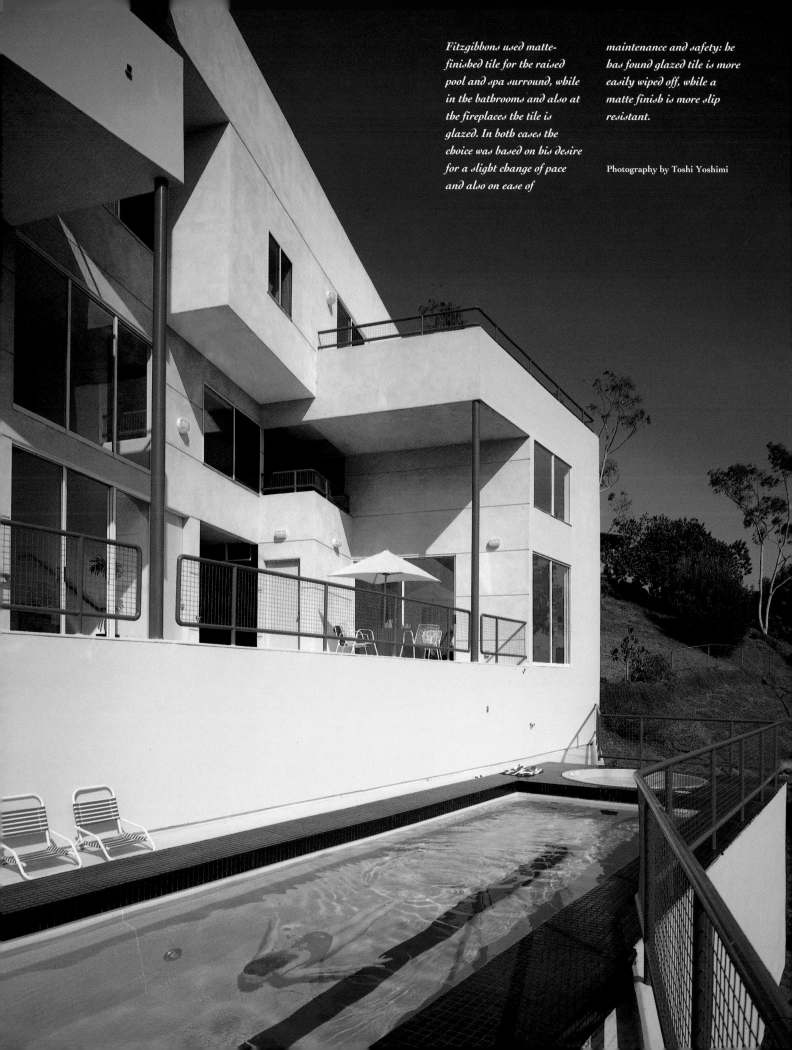

Fitzgibbons used matte-finished tile for the raised pool and spa surround, while in the bathrooms and also at the fireplaces the tile is glazed. In both cases the choice was based on his desire for a slight change of pace and also on ease of maintenance and safety: he has found glazed tile is more easily wiped off, while a matte finish is more slip resistant.

Photography by Toshi Yoshimi

UNLIMITED POSSIBILITIES
Syndecrete®

A NUMBER OF PRODUCTS COMING ON THE market now reflect an awareness of the need to conserve environmental resources. As an alternative to quarried or found stone, architect ***David Hertz*** developed Syndecrete®. Manufactured by Syndesis, his multidisciplinary architectural/design/ manufacturing firm, Syndecrete® is a

■ ■ ■

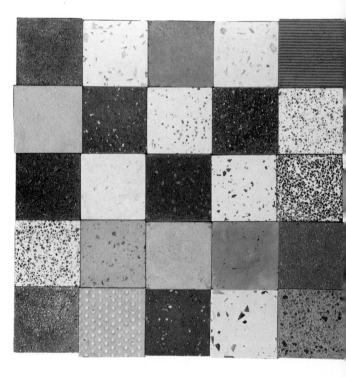

ABOVE: *Two-inch slabs of lightweight precast concrete cover existing ceramic tile seamlessly. The glaze has been set into precast grooves and is frameless.*

ABOVE RIGHT: *Custom aggregate mixes with precast lightweight concrete product include copper wire, recycled plastic, tempered glass, acrylic scraps. . . . Surface textures from bubble wrap to corrugated and wood grain to glass are possible. Designed by David Hertz, Syndesis, Inc.*

RIGHT: *Kitchen counter slabs made of Syndecrete®, designed by David Hertz, are typically two inches thick and solid throughout, minimizing plywood subtops required for tile and stone.*

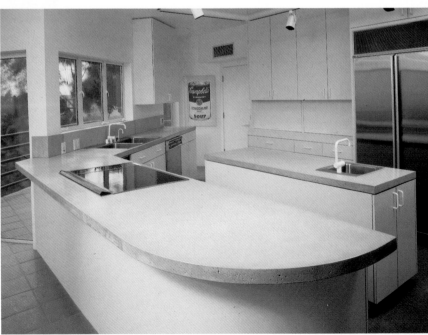

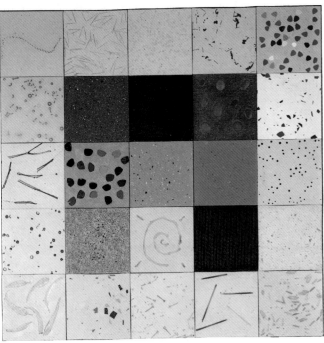

ABOVE LEFT: *Powder room vanity extends up the wall, eliminating the need for a back splash and providing a vertical surface for the attached mirror. Designed by David Hertz and Susan Stringfellow.*

LEFT: *The broad range of recycled materials used as Syndecrete® decorative aggregates includes nuts and bolts, wood chips, broken records, marbles, computer chips, and even plastic toy soldiers.*

ABOVE RIGHT: *This curved bar top, designed by Daryl Shapiro, was made without facing or seams, and without creating waste typical of slab or sheet products.*

Photography by Tom Bonner and Della Ventura (left)

RIGHT: *Mounting of conventional sinks and appliances is possible. Here the edge of the sink has been made only one inch thick.*

RIGHT: *Solid monolithic tables of cast Syndecrete®, designed by David Hertz.*

precast lightweight concrete architectural surfacing material handcrafted to meet individual specifications.

Offered only in precast pieces, Syndecrete® is not available as a dry mix and cannot be poured in place. Rather, each Syndecrete® piece is custom formed, poured and hand-finished at Hertz's Santa Monica, California, facility.

Custom-mix designs incorporating aggregates offer unlimited possibilities — recycling of post-consumer and scrap materials such as reground plastic, wood chips, crushed glass, metal shavings, stone fragments. . . . Surfaces can be ground, polished or textured to expose the natural porosity and aggregates in the Italian tradition of terrazzo.

The possibilities seem endless. As with the entire range of tile, stone and other

RIGHT: *Slight color variations give these twenty-four-inch-square, one-and-a-half-inch-thick Syndecrete® tiles an organic quality. Surfaces are ground smooth, sealed and waxed to give a warm, smooth surface. Designed by Mark Kirkhart and Dayne Van Kleek.*

LEFT: *Custom color chips of Syndecrete®. Integral pigmentation allows for a wide choice of colors.*

Photography by Tom Bonner and Rachel Olsen (above)

alternatives presented in this book, the final design truly depends on nothing but one's imagination. If that can be wedded with an awareness of what we are doing to our Planet Earth, then maybe she will continue being our Spaceship Earth for many years to come.

GLOSSARY

CERAMIC TILE

AMERICAN NATIONAL STANDARD SPECIFICATIONS FOR CERAMIC TILE (ANSI A137.1-1988)

Published by Tile Council of America, Inc.,
   P.O. Box 326
   Princeton, New Jersey 08542-0326
   Tel: (609) 921-7050   Fax: (609) 452-7255

Basis for Acceptance The method of determining whether a lot of ceramic tile is acceptable under these specifications.

Ceramic Mosaic Tile Tile formed by either the dust-pressed or plastic method, usually 1/4 to 3/8 inch thick, and having a facial area of less than six square inches. Ceramic mosaic tile may be of either porcelain or natural clay composition and may be either plain or with an abrasive mixture throughout.

Ceramic Tile See Tile.

Conductive Tile Tile made from special body composition or by methods that result in specific properties of electrical conductivity while retaining other normal physical properties of tile.

Decorative Thin Wall Tile A glazed tile with a thin body that is usually non-vitreous and that is suitable for interior decorative residential wall use where breaking strength is not a requirement.

Edgebonded Tile See Pregrouted Tile.

Facial Defect The portion of the facial surface of the tile which is readily observed to be nonconforming and which will detract from the aesthetic appearance or serviceability of the installed tile.

Glazed Tile Tile with a fused impervious facial finish composed of ceramic materials, fused to the body of the tile which may be non-vitreous, semi-vitreous, vitreous, or impervious.

Impervious Tile Tile with water absorption of 0.5 percent or less.

Module Size The actual tile dimension plus the manufacturer's recommended joint width. The module dimension is measured from center to center of the joints.

Mounted Tile Tile assembled into units or sheets by suitable material to facilitate handling and installation. Tile may be face-mounted, back-mounted or edge-mounted. Face-mounted tile assemblies may have paper or other suitable material applied to the face of each tile, usually by water soluble adhesives so that it can be easily removed after installation but prior to grouting of the joints. Back-mounted tile assemblies may have perforated paper, fiber mesh, resin or other suitable material bonded to the back and/or edges of each tile which becomes an integral part of the tile installation. Back-mounted and edge-mounted tile assemblies shall have a sufficient exposure of tile and joints surrounding each tile to comply with bond strength requirements. Tile manufacturers must specify whether back-mounted or edge-mounted tile assemblies are suitable for installation in swimming pools, on exteriors and/or in wet areas.

Natural Clay Tile A ceramic mosaic tile or a paver tile made by either the dust-pressed or the plastic method from clays that produce a dense body having a distinctive slightly textured appearance.

Nominal Sizes This is the approximate facial size or thickness of tile, expressed in inches or fractions of an inch, for general reference.

Non-Vitreous Tile Tile with water absorption of more than 7.0 percent.

Paver Tile Glazed or unglazed porcelain or natural clay tile formed by the dust-pressed method having a facial area of six square inches or more.

Physical Properties of Ceramic Tile Those properties as measured by the ASTM tests referred to herein. Copies of the ASTM test procedures may be obtained from:
   American Society for Testing and Materials
   1916 Race Street
   Philadelphia, Pennsylvania 19103
When no standard test is available from ASTM the test procedure required is included in 10.1.1, 10.1.2 and 10.1.3.

Porcelain Tile A ceramic mosaic tile or paver tile that is generally made by the dust-pressed method of a composition resulting in a tile that is dense, impervious, fine grained, and smooth with sharply formed face.

Pregrouted Tile A surfacing unit consisting of an assembly of ceramic tiles bonded together at their edges by a material, generally elastomeric, which seals the joints completely. Such material (grout) may fill the joint completely, or partially and may cover all, a portion or none of the back surfaces of the tiles in the sheets. The perimeter of these factory pregrouted sheets may include the entire, or part of the joint between the sheets or none at all. The term Edgebonded Tile is sometimes used to designate a particular type of pregrouted tile sheets having the front and back surfaces completely exposed.

Quarry Tile Glazed or unglazed tile, made by the extrusion process from natural clay or shale usually having a facial area of six square inches (39 cm²) or more.

Sampling The method of obtaining tile for testing from an agreed-upon lot.

Self-Spacing Tile Tile with lugs, spacers, or protuberances on the sides which automatically space the tile for grout joints.

Semi-Vitreous Tile Tile with water absorption of more than 3.0 percent, but not more than 7.0 percent.

Slip-Resistant Tile Tile usually having greater slip-resistant characteristics due to an abrasive admixture, abrasive particles in the surface, grooves or patterns in the surface or a glaze specifically designed for increased coefficient of friction.

Special Purpose Tile Tile, either glazed or unglazed, made to meet or to have special physical design or appearance characteristics such as size, thickness, shape, color, or decoration; keys or lugs on backs or sides; pregrouted assemblies or sheets; special resistance to staining, frost, alkalies, acids, thermal shock, physical impact, or high coefficient of friction.

Structural Defects Cracks or laminations in the body of the tile which detract from the aesthetic appearance and/or the structural soundness of the tile installation.

System Modularity Tiles of various nominal dimensions are sized so that they may be installed together in patterns with a common specified joint width.

Testing of Ceramic Tile The act of determining whether ceramic tile is acceptable; see Physical Properties of Ceramic Tile.

Tile A ceramic surfacing unit, usually relatively thin in relation to facial area, made from clay or a mixture of clay and other ceramic materials, called the body of the tile, having either a glazed or unglazed face and fired above red heat in the course of manufacture to a temperature sufficiently high to produce specific physical properties and characteristics.

Tile Assemblies See Mounted Tile.

Trim Units Units of various shapes consisting of such items as bases, caps, corners, moldings, angles, etc., necessary to achieve an installation of the desired sanitary and/or architectural design.

Unglazed Tile A hard, dense tile of uniform composition throughout, deriving color and texture from the materials of which the body is made.

Vitreous Tile Tile with water absorption of more than 0.5 percent, but not more than 3.0 percent.

Wall Tile A glazed tile with a body that is suitable for interior use and which is usuall non-vitreous, and is not required or expected to withstand excessive impact or be sub ject to freezing and thawing conditions.

Wet Area Tile surfaces that are either soaked, saturated, or subjected to moisture o liquids (usually water) such as gang showers, tub enclosures, showers, laundries, sauna steamrooms, swimming pools and exterior areas

STONE

Edited by **Robert Hund,** *Managing Director*
Marble Institute of America, Inc.
33505 State Street
Farmington, Michigan 48335
Tel: (810) 476-5558   Fax: (810) 476-1630

Abrasive Finish (Sand Rubbed Finish) A flat, nonreflective surface finish.

Abrasive Hardness (Ha) The wearing qualities of stone for floors, stair treads, and other areas subjected to abrasion by foot traffic.

Absorption Percentage of moisture absorption by weight.

Adoquin A volcanic quartz-based stone containing a variety of colored aggregates and pumice in a quartz matrix. Quarried in Mexico. Available in several colors.

Agglomerated "Stone" A product made from quarry waste.

Aggregate
1. Quantities of loose fragments of rock or mineral.
2. The sand or gravel that when added to cement and water makes concrete. Aggregate is described as fine (sand) or coarse (gravel).

Alabaster Fine-grained, translucent variety of gypsum, generally white in color. May be cut or carved easily with a knife or saw. Term is often incorrectly applied to fine-grained marble.

Anchors Metal (mechanical) devices for securing dimension stone units to structural members or back-up walls.
Types for stonework:
1. Flat Stock—Strap, cramps, dovetails, strap and dowel, and two-way anchors.
2. Corrugated—Corrugated wall ties and dovetail anchors.
3. Round Stock—Rod cramp, rod anchor, dowel, eyebolt and dowel, dowel and wire toggle bolts, and wire.

Antiqued Marble See Tumbled Marble.

Basalt A dark-colored, densely textured igneous rock relatively high in iron and magnesia minerals and relatively low in silica, commercially known as granite when fabricated as dimension stone.

Black Granite Dark-colored igneous rocks (defined by geologists as basalt, diabase, gabbro, diorite, and anorthosite) quarried as building stone, building facings, and specialty purposes and identified as Black Granite when sold.

Bluestone A fine- to medium-grained, hard, metamorphic quartz-based stone of characteristic blue, gray, green and buff colors quarried in the U.S. Appalachian Plateau and in other countries. Much bluestone splits readily along original bedding planes to form thin slabs. Formed in the Devonian Age, the Upper Stone is green and lilac in color, while the Middle Stone is dark gray and blue.

Book Matching Veneer slabs cut and assembled so that one slab will match the other in the horizontal direction, or in a vertical direction, but not both. Slabs must have alternate faces finished in sequence as they are layered in the quarry block.

Breccia Stone in which angular fragments are imbedded in a matrix of the same or another composition.

Brownstone A dark brown or reddish-brown arkosic quartz-based stone (sandstone), extensively used for construction in the U.S. during the nineteenth century. Characteristic brown or reddish-brown color is due to the presence of iron oxide minerals as bonding or interstitial material. Stone for New York City's noted "brownstone fronts" came from the Connecticut Valley in Massachusetts, southeastern Pennsylvania and New Jersey.

Brushed Finish Obtained by brushing the stone with a coarse rotary-type wire brush.

Building Stone Natural rock of adequate quality to be quarried and cut as dimension stone as it exists in nature.

Bush-Hammered Finish A textured, tooled finish produced by a mechanical process; the texture varies from subtle to rough.

Calcite A crystalline variety of limestone containing not more than 5 percent magnesium carbonate.

Cobblestone A naturally rounded dimension stone, large enough for use in paving. Also, a term commonly used to describe paving blocks (usually granite), generally cut to rectangular shapes.

Commercial Definition of Marble A crystalline rock, capable of taking a polish, and composed of one or more of three minerals—calcite, dolomite, or serpentine.

Coquina Limestone composed predominantly of shells or fragments of shells loosely cemented by calcite. Coquina is coarse-textured and has a high porosity. The term is applied principally to a very porous rock quarried in Florida.

Cultured Marble An artificial, man-made product, created by mixing minimal amounts of marble dust into a resin.

Diabase A granular igneous rock, dark gray to black, sometimes called dolerite.

Diamond Matching (Quarter Matching) A veneer panel matching pattern similar to book matching, except that the third and fourth panels are inverted over panels one and two.

Dimension Stone Natural building stone that has been selected, trimmed, and cut to specified shapes and sizes, with minimal size tolerances, and with or without surface finishing.

Dolomitic Limestone A limestone rich in magnesium carbonate, frequently somewhat crystalline in character. It is found in ledge formations in a wide variety of color tones and textures. Generally speaking, its crushing and tensile strengths are greater than oolitic limestones, and its appearance shows greater variety in texture.

Efflorescence A crystalline deposit with the appearance of whitish powder which occurs on stone surfaces and is caused by the deposition of soluble salts carried through or onto the surface by moisture.

Flagstone Thin slabs of stone with one moderately smooth surface and often irregular shaped bottom, used for paving walks, driveways, patios, etc. They are generally fine-grained bluestone, other quartz-based stone or slate. Thin slabs of other stones may also be used.

Flamed Finish See Thermal Finish.

Fleuri Cut The mottled effect obtained when certain stone varieties are cut parallel to their natural bedding plane.

Gabbro An igneous granular stone composed chiefly of pyroxene, augite or diallage, and plagioclase.

Gneiss A foliated crystalline rock composed essentially of silicate minerals with interlocking and visibly granular texture in which the foliation is due primarily to alternating layers, regular or irregular, of contrasting mineralogic composition.

Granite A very hard, crystalline, igneous rock, gray to pink in color, composed of feldspar, quartz, and lesser amounts of dark ferromagnesium materials. Gneiss and black "granites" are similar to "true" granites in structure and texture, but are composed of different minerals.

Greenstone A metamorphic rock, typically with poorly defined granularity, ranging in color from medium green or yellowish green to black, owing to the presence of chlorite, epidote, or actinolite minerals.

Honed Finish Surface is ground to a satin-smooth, super-fine finish having little or no gloss.

Igneous One of the three great classes of rock—igneous, sedimentary and metamorphic—solidified from molten slate, as granite and lava.

Kerf Slot cut into the edge of stone with saw blade for insertion of anchors.

Limestone A sedimentary rock composed primarily of calcite or dolomite. The varieties of limestone used as dimension stone are usually well consolidated and exhibit a minimum of graining or bedding direction. Limestones that contain not more than 5 percent magnesium carbonate may be termed calcite limestone, as distinguished from those that contain between 5 and 40 percent magnesium carbonate (magnesium or dolomitic limestone), and from those that contain in excess of 40 percent as the mineral dolomite (dolostone, formerly known as the rock dolomite). Recrystallized limestones and compact, dense, relatively pure microcrystalline varieties that are capable of taking a polish are known commercially as marbles.

Lippage A condition where one edge of a stone is higher than adjacent edges, giving the finishes' surface a ragged appearance.

Marble A crystalline rock composed predominantly of crystalline grains of calcite, dolomite or serpentine, and capable of taking a mechanical polish. Marble is generally a recrystallization of limestone.

Matching Selecting, cutting and placing finished stone slabs to obtain a uniform and symmetrical pattern of natural veining and color. (See: Book Matching, Diamond Matching and Slip Matching.)

Metamorphic Rocks altered in appearance, density and crystalline structure from metamorphism. Slate is derived from shale, quartz-based stone from quartzitic sand, and marble from limestone.

Metamorphism The change or alteration in a rock caused by exterior agencies, such as deep-seated heat and intense pressure, or intrusion of rock materials.

Onyx A generally translucent and usually layered cryptocrystalline calcite stone with colors in pastel shades, particularly yellow, tan and green. Often marketed as a type of marble, although it is not technically a marble.

Oolithic Limestone A limestone composed largely of spherical or subspherical particles called oolites or ooliths; a calcite-cemented calcareous stone formed of shells and shell fragments, practically non-crystalline in character. It is found in massive deposits located almost entirely in Lawrence, Monroe and Owen counties, Indiana, and in Alabama, Kansas and Texas. This limestone is characteristically a freestone, without cleavage planes, possessing a remarkable uniformity in composition, texture and structure. It possesses a high internal elasticity, adapting itself without damage to extreme temperature changes.

Paver (Paving Stone) A single unit of fabricated stone for use as an exterior paving material.

Polished Finish The smoothest and glossiest surface available, which brings out the fu color, reflective property, and character of the stone. Used with marble, granite, onyx travertine and some limestones. This finish is not recommended for floors, as traffic wi eventually create dull-finish paths.

Quartz-Based Stone This stone may be either sedimentary in formation (as in sand stone) or metamorphic (as in quartzite).

Quartzite A compact granular quartz-based stone composed of quartz crystals, usually so firmly cemented as to make the mass homogenous. Quartzite is a highly indurated typically metamorphosed stone containing at least 95 percent free silica, which fracture conchoidally through the grains. The stone is generally quarried in stratified layers, the surfaces of which are usually somewhat smooth. Its crushing and tensile strengths are extremely high.

Rough Sawn Finish A surface finish resulting from the gang sawing process.

Rubbed Finish Mechanically rubbed for smoother finish. May have slight "trails" o scratches.

Sandblasted, Coarse-Stippled Coarse plane surface produced by blasting with a abrasive; coarseness varies with type of preparatory finish and grain structure of the stone

Sandblasted, Fine-Stippled Plane surface, slightly pebbled, with occasional slight "trails or scratches.

Sandstone A sedimentary rock consisting usually of sand-sized (2-0.06 mm) quartz fragments, having a minimum of 60 percent free silica, and cemented or bonded together b one of the interstitial or bonding materials by which sandstones are classified.

Scagliola
1. Plasterwork used in imitation of ornamental marble, consisting of ground gypsum and glue colored with marble or granite dust.
2. A small piece of marble.

Schist A foliated metamorphic rock. The more common schists are composed of mi cas and generally contain subordinate quartz and/or feldspar.

Sedimentary Rocks formed of sediments laid down in successive strata or layers. The materials of which they are formed are derived from preexisting rocks or the skeletal remains of sea creatures.

Serpentine (Serpentine Marble) A stone consisting primarily or entirely of serpentine (magnesium silicate); generally dark green in color with markings of white, light green or black. Considered commercially as a marble because it can be polished. One of the hardest varieties of natural building stone.

Setter An experienced journeyman who installs dimension stone.

Setting The trade of installing finished dimension stone.

Slate A very fine-grained metamorphic rock derived from sedimentary rock shale and composed mostly of micas, chlorite and quartz. Characterized by an excellent parallel cleavage entirely independent of original bedding. Slate is easily split along the cleavage into relatively thin but strong slabs.

Slip Matching Veneer panels all finished on the same face and placed side by side, forming a repetition of the same pattern in each panel.

Soapstone A massive variety of talc with a "soapy" feel used for hearths, tabletops chemical-resistant laboratory tops, stove facings and cladding; known for its stain-proof characteristics.

Soundness A property of stone used to describe relative freedom from cracks, faults voids and similar imperfections found in untreated stone. One of the characteristics en-

Stone *(Continued)*

countered in fabrication. Marble and Limestone Marble have been classified into groups: A, B, C, and D. (See "Recommendations for Commercial Floor Marble" at the end of this glossary.*)

Sticking A trade term describing the butt edge repair of a broken piece of stone, now generally done with dowels, cements or epoxies.

Syenite Granite-like rock containing little or no quartz.

Terrazzo A flooring surface in which marble or granite chips are mixed into a cementitious or resinous matrix, then ground to a flat surface, exposing the chips. The exposed chips are then finished and take a high polish.

Thermal Finish A rough finish created with the use of intense heat flaming to exfoliate the surface of the stone and expose the actual grain. Large surfaces may have shadow lines caused by overlapping of the flaming machine or torch.

Thin Stone Dimension units less than two inches thick.

Translucence The ability of many lighter colored marbles and onyx to transmit light.

Travertine A variety of crystalline or microcrystalline limestone distinguished by a laystructure, and with pores and cavities concentrated in some of the layers, creating an open texture. Many varieties of travertine take a polish and are known commercially as marble.

Tumbled Marble A special finish for marble, limestone marble and limestone obtained by rotating pre-cut pieces in a mixer or other container. This rounds the edges and arrises.

Veneer An interior or exterior thin stone, two inches or less in thickness, used as the exposed, decorative wall facing material; not intended to be load-bearing.

Verde Antique A marble composed chiefly of massive serpentine and capable of being polished. It is commonly crossed by small veins of other minerals, chiefly carbonates of calcium and magnesium.

Waxing A stone trade expression referring to the practice of filling minor surface imperfections such as voids or sand holes with melted shellac, cabinetmaker's wax or certain polyester compounds. It does not refer to the application of paste wax to make surfaces shinier.

RECOMMENDATIONS FOR COMMERCIAL FLOOR MARBLE

Prepared by the **Marble Institute of America, Inc.**

1. Minimum ³/₄-inch thickness.
2. A honed finish.
3. A "hardness" value of a minimum "12" as measured by ASTM (American Society for Testing and Materials) C241.
4. A Soundness Classification of "A"—although marbles with lesser Soundness can be considered—if waxing, sticking, filling, cementing and reinforcing are properly done.

Soundness

As a result of knowledge gained in extensive practical experience of the members of the Marble Institute of America, marbles have been classified into four groups. The basis of this classification is the characteristics encountered in fabricating and has no reference whatsoever to comparative merit or value. The classifications merely indicate what method of fabrication is considered necessary and acceptable in each instance, as based on standard trade practice. Classification of marble is done by MIA producer and finisher members. A written warranty should be obtained from them prior to installation.

The groupings—**A**,**B**,**C**, and **D**—should be taken into account when specifying marble, for all marbles are not suitable for all building applications. This is particularly true of the comparatively fragile marbles classified under groups C and D which may require additional fabrication before or during installation. These four groups are:

Group A
Sound marbles with uniform and favorable working qualities; containing no geological flaws or voids.

Group B
Marbles similar in character to the preceding group; but with less favorable working qualities; may have natural faults; a limited amount of waxing, sticking and filling may be required.

Group C
Marbles with some variations in working qualities; geological flaws, voids, veins and lines of separation are common. It is standard practice to repair these variations by one or more of several methods—waxing, sticking, filling or cementing. Liners and other forms of reinforcement are used when necessary.

Group D
Marbles similar to the preceding group, but containing larger proportion of natural faults, maximum variations in working qualities, and requiring more of the same methods of finishing. This group comprises many of the highly colored marbles prized for their decorative values.

The Soundness Classifications merely indicate what method and amount of repair and fabrication is necessary prior to or during installation, as based on standard trade practices.

BRICK

published by **Brick Institute of America**
11490 Commerce Park Drive
Reston, Virginia 22091
Tel: (703) 620-1200   Fax: (703) 620-3928
Edited from *Glossary of Terms Relating to Brick Masonry*

Brick A solid masonry unit of clay or shale, formed into a rectangular prism while plastic and burned or fired in a kiln.

Acid-Resistant Brick Brick suitable for use in contact with chemicals, usually in conjunction with acid-resistant mortars.

Adobe Brick Largely roughly molded, sun-dried clay brick of varying size.

Angle Brick Any brick shaped to an oblique angle to fit a salient corner.

Arch Brick
1. Wedge-shaped brick for special use in an arch.
2. Extremely hard-burned brick from an arch of a scove kiln.

Building Brick Brick for building purposes not specially treated for texture or color. Formerly called *common brick*.

Clinker Brick A very hard-burned brick whose shape is distorted or bloated due to nearly complete vitrification.

Common Brick See Building Brick.

Dry-Press Brick Brick formed in molds under high pressures from relatively dry clay (5 to 7 percent moisture content).

Economy Brick Brick whose nominal dimensions are 4 by 4 by 8 inches.

Engineered Brick Brick whose nominal dimensions are 4 by 3^1/s by 8 inches.

Facing Brick Brick made especially for facing purposes, often treated to produce surface texture. They are made of selected clays, or treated, to produce desired color.

Fire Brick Brick made of refractory ceramic material which will resist high temperatures.

Floor Brick Smooth dense brick, highly resistant to abrasion, used as finished floor surfaces.

Gauged Brick
1. Brick which has been ground or otherwise produced to accurate dimensions.
2. A tapered arch brick.

Hollow Brick A masonry unit of clay or shale whose net cross-sectional area in any plane parallel to the bearing surface is not less than 60 percent of its gross cross-sectional area measured in the same plane.

Jumbo Brick A generic term indicating a brick larger in size than the standard. Some producers use this term to describe oversize brick of specific dimensions manufactured by them.

Norman Brick A brick whose nominal dimensions are 4 by 2^2/3 by 12 inches.

Paving Brick Vitrified brick especially suitable for use in pavements where resistance to abrasion is important.

Roman Brick Brick whose nominal dimensions are 4 by 2 by 12 inches.

Salmon Brick Generic term for underburned bricks that are more porous, slightly larger, and lighter colored than hard-burned brick. Usually pinkish-orange color.

Soft-Mud Brick Brick produced by molding relatively wet clay (20 to 30 percent moisture). Often a hand process. When insides of molds are sanded to prevent sticking of clay, the product is *sand-struck* brick. When molds are wetted to prevent sticking, the product is *water-struck* brick.

Stiff-Mud Brick Brick produced by extruding a stiff but plastic clay (12 to 15 percent moisture) through a die.

Ceramic Color Glaze An opaque colored glaze of satin or gloss finish obtained by spraying the clay body with a compound of metallic oxides, chemicals and clays. It is burned at high temperatures, fusing glaze to body, making them inseparable.

Clay A natural, mineral aggregate consisting essentially of hydrous aluminum silicate; is plastic when sufficiently wetted, rigid when dried, and vitrified when fired to a sufficiently high temperature.

Efflorescence A powder or stain sometimes found on the surface of masonry, resulting from deposition of water-soluble salts.

Engineered Brick Masonry Masonry in which design is based on a rational structural analysis.

Face
1. The exposed surface of a wall or masonry unit.
2. The surface of a unit designed to be exposed in the finished masonry.

Facing Any material, forming a part of a wall, used as a finished surface.

Fire Clay A clay which is highly resistant to heat without deforming and used for making brick.

Grout Mixture of cementitious material and aggregate to which sufficient water is added to produce pouring consistency without segregation of the constituents.

Kiln A furnace oven or heated enclosure used for burning or firing brick or other clay material.

Masonry Cement A mill-mixed cementitious material to which sand and water must be added.

Mortar A plastic mixture of cementitious materials, fine aggregate and water.

Prefabricated Brick Masonry Masonry construction fabricated in a location other than its final inservice location in the structure. Also known as preassembled, panelized and sectionalized brick masonry.

Shale Clay that has been subjected to high pressures until it has hardened.

Soft-Burned Clay products that have been fired at low temperature ranges, producing relatively high absorptions and low compressive strengths.

DIRECTORY

ARCHITECTS & DESIGNERS

Reginald Adams
Reginald Adams & Associates
8500 Melrose Avenue, Suite 207
Los Angeles, California 90069
United States
Tel: (310) 659-8038
Fax: (310) 659-8594

Ace Architects
   Lucia Howard
   David Weingarten
The Leviathan
330 Second Street, #1
Oakland, California 94607
United States
Tel: (510) 452-0775
Fax: (510) 452-1175

Joseph Alcasar Terrell
Alcasar Terrell, Inc.
119½ North Larchmont Boulevard
Los Angeles, California 90004
United States
Tel: (213) 469-8044
Fax: (213) 469-0561

Joaquin Alvarez Montes
A. M. Associates International
Madrid 28010
Spain

AM Partners, Inc.
   Charles K. C. Lau, AIA, ISP
   Lois Zanteson
1164 Bishop Street, Suite 1000
Honolulu, Hawaii 96813
United States
Tel: (808) 526-2828
Fax: (808) 538-0027

Penny Drue Baird
Dessins Inc.
285 Lafayette Street
New York, New York 10012
United States
Tel: (212) 431-1380
Fax: (212) 431-1608

Samuel Botero
Samuel Botero Associates, Inc.
150 East 58th Street, 23rd Floor
New York, New York 10155
United States
Tel: (212) 935-5155
Fax: (212) 832-0714

Barry Brukoff
Brukoff Design Associates, Inc.
480 Gate Five Road, Suite 310
Sausalito, California 94965
United States
Tel: (415) 332-6350
Fax: (415) 332-5968

Erika Brunson
Erika Brunson Design Associates
903 Westbourne Drive
Los Angeles, California 90069
United States
Tel: (310) 652-1970
Fax: (310) 652-2381

Arthur de Mattos Casas
Rua Manuel Maria Tourinho, 46
01236 000 São Paulo, São Paulo
Brazil
Tel: (55) 11 282 6311   (55) 11 62 0243
Fax: (55) 11 282 6608

Centerbrook Architects
   Mark Simon, FAIA
   James C. Childress, AIA
P.O. Box 955
Essex, Connecticut 06426
United States
Tel: (203) 767-0175
Fax: (203) 767-8719

Theodore Ceraldi
Theodore Ceraldi & Associates
P.O. Box 13
Nyack, New York 10960
United States
Tel: (914) 353-1199
Fax: (914) 353-1314

Alfredo De Vido, FAIA
Alfredo De Vido Associates
1044 Madison Avenue
New York, New York 10021
United States
Tel: (212) 517-6100
Fax: (212) 517-6103

Steven Ehrlich
Steven Ehrlich Architects
2210 Colorado Avenue
Santa Monica, California 90404
United States
Tel: (310) 828-6700
Fax: (310) 828-7710

Rand Elliott, FAIA
Elliott + Associates Architects
6709 North Classen, Suite 101
Oklahoma City, Oklahoma 73116
United States
Tel: (405) 843-9554
Fax: (405) 843-9607

Nury Feria, ASID, IDG
Design Perceptions, Inc.
139 North East 40th Street, Suite 203
Miami, Florida 33137
United States
Tel: (305) 573-3128
Fax: (305) 576-9257

J. Frank Fitzgibbons, AIA
Fitzgibbons Associates Architects
4822 Glencairn Road
Los Angeles, California 90027
United States
Tel: (213) 663-7579
Fax: (213) 663-6262

Ford, Powell & Carson
   Chris Carson, FAIA
   John Gutzler, ASID
1138 East Commerce Street
San Antonio, Texas 78205
United States
Tel: (210) 226-1246
Fax: (210) 226-6482

Jaime Gesundheit
Gesundheit Architects
14951 Califa Street
Van Nuys, California 91411
United States
Tel: (818) 781-1390
Fax: (818) 781-1467

Marisabel Gomez de Morales
Arquitectura de Interiores
Aurelio Ortega - #764-D
Guadalajara
Jalisco
Mexico 45150
Tel: (523) 656 2939
Fax: (523) 656 5747

Gomez Vazquez Aldana & Associates
   Jaime and J. Manuel Gomez Vazquez
Aurelio Ortega - #764
Guadalajara
Jalisco
Mexico 45150
Tel: (523) 656 4343
Fax: (523) 656 4087

Donald C. Hensman, FAIA
Buff, Smith & Hensman
1450 West Colorado Boulevard
Pasadena, California 91105
United States
Tel: (818) 795-6464
Fax: (818) 795-0961

David Hertz, AIA
Syndesis, Inc.
2908 Colorado Avenue
Santa Monica, California 90404
United States
Tel: (310) 829-9932
Fax: (310) 829-5641

George V. Hogan, AIA
Hogan & Chapman
263 A Kaelepulu Drive
Kailua, Hawaii 96734
United States
Tel: (808) 263-4656

Allison A. Holland, ASID
Creative Decorating
168 Poloke Place
Honolulu, Hawaii 96822
United States
Tel: (808) 955-1465
Fax: (808) 949-2290

Dennis Jenkins
Dennis Jenkins Associates
5813 South West 68th Street
South Miami, Florida 33143
United States
Tel: (305) 665-6960
Fax: (305) 665-6971

James F. Jereb, Ph.D.
Tribal Design
1001 East Alameda
Santa Fe, New Mexico 87501
United States
Tel: (505) 989-8765
Fax: (505) 989-3353

Tessa Kennedy, IIDA
Tessa Kennedy Design, Ltd.
91/97 Freston Road
London W11 4BD
England
Tel: (44) 71 221 4546
Fax: (44) 71 229 2899

Renée Kubiak, IIDA
P.O. Box 665
Huntington Beach, California 92648
United States
Tel/Fax: (714) 960-3072

Michael La Rocca
150 East 58th Street, Suite 3510
New York, New York 10015
United States
Tel: (212) 755-5558
Fax: (212) 838-3034

William Leddy, AIA
Tanner Leddy Maytum Stacy Architects
444 Spear Street
San Francisco, California 94105
United States
Tel: (415) 394-5400
Fax: (415) 394-5400

Sally Sirkin Lewis
Sally Sirkin Lewis Interior Design
8727 Melrose Avenue
Los Angeles, California 90069
United States
Tel: (310) 659-4910
Fax: (310) 859-8935

Heljä Liukko-Sundström
Marjatantie 24
00610 Helsinki
Finland
Tel: (358) 0 39391
Fax: (358) 0 791 329

Irene Montgomery, ASID
Box 7000-339
1719 Via el Prado
Redondo Beach, California 90277
United States
Tel/Fax: (310) 373-7567

Juan Montoya
Juan Montoya Design Corporation
80 Fifth Avenue
New York, New York 10011
United States
Tel: (212) 242-3622
Fax: (212) 242-3743

Ramon Pacheco
Ramon Pacheco & Associates
4990 South West 72nd Avenue, Suite 101
Miami, Florida 33155
United States
Tel: (305) 666-2573
Fax: (305) 666-3871

Pascal Arquitectos
    Carlos Pascal and Gerard Pascal
Atlaltunco - #99
Tecamachalco
Mexico
Tel: (525) 294 2371
Fax: (525) 294 8513

John Portman
John Portman & Associates
231 Peachtree Street, Suite 200
Atlanta, Georgia 30303
United States
Tel: (404) 614-5555
Fax: (404) 614-5553

Powell/Kleinschmidt
    Donald Powell
    Robert D. Kleinschmidt
645 North Michigan Avenue
Chicago, Illinois 60611
United States
Tel: (312) 642-6450
Fax: (312) 642-5135

Vishva Priya
Ahuja Priya Architects
561 Broadway
New York, New York 10012
United States
Tel: (212) 219-2122
Fax: (212) 941-1456

Samuel Sandler
Madero 314 PTE
Monterrey, N.L.
64000 Mexico
Tel: (528) 374 48 54
Fax: (528) 375 30 83

Sheri Schlesinger
Schlesinger & Associates
101 South Robertson Boulevard, #202
Los Angeles, California 90048
United States
Tel: (310) 275-1330
Fax: (310) 275-8698

Louis Shuster
Shuster Design Associates, Inc.
1401 East Broward Boulevard, Suite 103
Fort Lauderdale, Florida 33301
United States
Tel: (305) 462-6400
Fax: (305) 462-6408

Martin Smyth
Steven J. Leach, Jr. + Associates
10/F Tai Sang Commercial Building
24-34 Hennessy Road
Hong Kong
Tel: (852) 528 5544
Fax: (852) 861 0354

Warren H. Snodgrass
Design Technology
2105 Shelter Bay Avenue
Mill Valley, California 94941
United States
Tel: (415) 381-2353
Fax: (415) 381-4245

German C. Sonntag, ASID
1303 Oak Street, Suite C
Santa Monica, California 90405
United States
Tel/Fax: (310) 452-2757

Richard Stowers
Stowers Associates
1978 The Alameda
San Jose, California 95126
United States
Tel: (408) 247-1416
Fax: (408) 247-1242

The System Design
    Mark Warwick
    Kim Hoffman
9828 Charleville Boulevard
Beverly Hills, California 90212
United States
Tel: (310) 556-7711
Fax: (310) 788-3808

Chris Tosdevin
Bulthaup
153 South Robertson Boulevard
Los Angeles, California 90048
United States
Tel: (310) 288-3875
Fax: (310) 288-3885

Studio Transit Design
    Giovanni Ascarelli, Maurizio Macciocchi
    Evaristo Nicolao and Danilo Parisio
Via Emilio Morosini, 17
00153 Roma
Italy
Tel: (39) 6 5899848
Fax: (39) 6 5898431

William F. Tull
Tull Company
5658 North Scottsdale Road
Scottsdale, Arizona 85253
United States
Tel: (602) 944-0575

Oscar Tusquets Blanca
Tusquets, Diaz & Associates
Cavallers, 50
08034 Barcelona
Spain
Tel: (34) 3 280 55 99
Fax: (34) 3 280 4071

Trisha Wilson
Wilson & Associates
3811 Turtle Creek Boulevard
Dallas, Texas 75219
United States
Tel: (214) 521-6753
Fax: (214) 521-0207

Wimberly Allison Tong & Goo, Inc.
    Gerald L. Allison, FAIA
    Eduardo Aguilar Robles, Arq.
    Bobby L. Caragay, AIA
2260 University Drive
Newport Beach, California 92660
United States
Tel: (714) 574-8500
Fax: (714) 574-8550

Alison Wright, AIA
Alison Wright Architects
8800 Venice Boulevard
Los Angeles, California 90034
United States
Tel: (310) 559-7467
Fax: (310) 559-2250

MANUFACTURERS, DISTRIBUTORS, SUPPLIERS, CONTRACTORS & ARTISANS

Adriatic Cast Stone
7402 Varna Avenue
North Hollywood, California 91605
United States
Tel: (818) 982-2666

American Olean Tile Co.
1000 Cannon Avenue
Lansdale, Pennsylvania 19446
United States
Tel: (215) 855-1111
Fax: (215) 393-2784

Arabia/Hackman Oy Ab
Hämeentie 135
00560 Helsinki
Finland
Tel: (358) 17 67 211
Fax: (358) 17 672 1283

Arte En Cantera
Avenue Lazaro Cardenas, #111
Plaza Brisas
Col. Valle Las Brisas
Monterrey N.L. 64780
Mexico
Tel: (528) 349 0140
Fax: (528) 357 9450

Asper Studio
Via Ditor Fiorenza 27
00199 Roma
Italy
Tel: (39) 6 831 0909
Fax: (39) 6 616 107

Banta Tile and Marble Company
P. O. Box 4032
Lancaster, Pennsylvania 17684
United States
Tel: (717) 393-3931

Bogart Masonry
P. O. Box 2398
Yucca Valley, California 92286
United States
Tel: (619) 364-2471

Juan Bordes
C/ Arapiles, 10
28015 Madrid
Spain
Tel: (34) 1 593 89 06
Fax: (34) 1 446 44 81

Bourget Bros. Coast Flagstone
1810 Colorado Avenue
Santa Monica, California 90404
United States
Tel: (310) 829-4010
Fax: (310) 829-6261

Byzantium
7255 South West 48th Street
Miami, Florida 33155
United States
Tel: (305) 669-1670
Fax: (305) 669-1839

Camasa
Calle 7 - #244 x 36
Frac. Campestre
Merida
Yucatan 97120
Mexico
Tel: (529) 923 9994
Fax: (529) 923 9994

J. F. Chen Antiques
8414 Melrose Avenue
Los Angeles, California 90069
United States
Tel: (213) 655-6310
Fax: (213) 655-9689

Concept Studios, Inc./Richard Goddard
2720 East Coast Highway
Corona Del Mar, California 92625
United States
Tel: (714) 459-0606

Connecticut Stone Supplies, Inc.
311 Boston Post Road
Orange, Connecticut 06477
United States
Tel: (203) 795-9767
Fax: (203) 799-9573

Country Floors, Inc.
15 East 16th Street
New York, New York 10003
Tel: (212) 627-8300
Fax: (212) 627-7742

Dennis & Leen
8734 Melrose Avenue
Los Angeles, California 90069
United States
Tel: (310) 652-0855
Fax: (310) 659-4311

Denny Moore
990 Ala Nanala 2-C
Honolulu, Hawaii 96818
United States
Tel/Fax: (808) 836-0048

Di Camillo Marble Accessories
Star Ridge Road
Brewster, New York 10509
United States
Tel: (914) 279-6571
Fax: (914) 424-3169

Domus Tiles
33 Pargate Road
London SW11 4NP
England
Tel: (44) 7122 35555

Doyle Construction
RFD 477B
Vineyard Haven, Massachusetts 02568
United States
Tel: (508) 693-9004
Fax: (508) 693-3221

Du Pont Co. - Corian®
Wilmington, Delaware 19898
United States
Tel: (800) 4-CORIAN
Fax: (800) 635-9496

Empire Marble & Granite, Inc.
2900 North West 77th Court
Miami, Florida 33122
United States
Tel: (305) 592-0029
Fax: (305) 591-4429

Eurocal Slate Centers
7741 Fay Avenue
La Jolla, California 92037
United States
Tel: (619) 551-9951
Fax: (619) 551-9951

Fachadas y Monumentos
2A De Magnolia 1615
Col. Reforma
Monterrey
Mexico
Tel: (528) 375 7770
Fax: (528) 375 0258

Formations
8746 Melrose Avenue
Los Angeles, California 90069
United States
Tel: (310) 659-3062
Fax: (310) 659-0694

Gina B.
8714 Santa Monica Boulevard
Los Angeles, California 90069
United States
Tel: (310) 652-4488
Fax: (310) 657-4180

Globe Marble & Tile, Inc.
7348 Bellaire Avenue
North Hollywood, California 91605
United States
Tel: (818) 982-4040

Hastings Tile Co.
30 Commercial Street
Freeport, New York 11520
United States
Tel: (516) 379-3500
Fax: (516) 379-3187

Impression
22599 South Western Avenue
Torrance, California 90501
Tel: (310) 618-1299
Fax: (310) 212-6719

Intercontinental Marble
8228 North West 56th Street
Miami, Florida 33166
United States
Tel: (305) 591-2207
Fax: (305) 477-3237

JB Marble Company
14654 Keswick Street
Van Nuys, California 91405
United States
Tel: (818) 902-1400
Fax: (818) 902-9681

Jeannee's Custom Tile
9525 Cozycroft Avenue, Unit K
Chatsworth, California 91311
Tel: (818) 709-1235

Walter Jenkins & Co. Ltd.
71 Upper Brockley Road
London SE4 1TF
England
Tel: (44) 8169 26655

Keystone Products, Inc.
1414 North West Third Avenue
Florida City, Florida 33034
United States
Tel: (305) 245-4716
Fax: (305) 245-8553

Latco Products
2943 Gleneden Street
Los Angeles, California 90039
United States
Tel: (213) 664-1171
Fax: (213) 665-6971

Marble Techniques
150 East 58th Street
New York, New York 10155
United States
Tel: (212) 750-9189
Fax: (212) 319-8349

Marmol y Arte
Av. Vasconcelos - #710 Poniente
Local 2
Garza Garcia, N.L. 66220
Monterrey N.L.
Mexico
Tel/Fax: (528) 338 9791

Ralph McIntosh
256 Naomi Avenue
Arcadia, California 91007
United States
Tel: (818) 446-9583
Fax: (818) 447-5667

McIntyre Tile
P.O. Box 14
Healdsburg, California 95448
United States
Tel: (707) 433-8866
Fax: (707) 433-0548

Sunny McLean
3800 North East Second Avenue
Miami, Florida 33137
United States
Tel: (305) 573-5943
Fax: (305) 573-1744

Midwest Marble & Granite Company
P.O. Box 1633
Manchester, Missouri 63011
United States
Tel: (314) 394-7227
Fax: (314) 527-0405

Andres Minero S. L.
Mar de Oman 14
Madrid 28033
Spain
Tel: (34) 1 314 6354

MOBAC, Inc.
220 Gale Lane, Suite 1
Kennett Square, Pennsylvania 19348
Tel: (610) 444-3490
Fax: (610) 444-4932

Angel Dilla Moline
General Yague 56
Madrid 28020
Spain
Tel: (34) 1 571 6440
Fax: (34) 1 571 6441

Thomas W. Morgan, Inc.
461 North Robertson Boulevard
Los Angeles, California 90048
United States
Tel: (310) 281-6450
Fax: (310) 281-6453

Natural Stone Ltd.
19345 North Indian Avenue
North Palm Springs, California 92258
United States
Tel/Fax: (619) 251-3277

Pacific Clay Products, Inc.
P.O. Box 549
Lake Elsinore, California 92531
United States
Tel: (909) 674-2131
Fax: (909) 674-4909

Reed Harris
Riverside House
Carnwath Road
London SW6 3HR
England
Tel: (44) 71 736 7511

Renaissance Marble
9116 De Garmo
Sun Valley, California 91352
United States
Tel: (818) 504-0100
Fax: (818) 504-2200

Rhomboid Sax Bath & Tile
8904 Beverly Boulevard
Los Angeles, California 90048
United States
Tel: (310) 550-0170
Fax: (310) 550-1803

Dominque Rocoffort
ROC Development
10642 Santa Monica Boulevard, Suite 206
Los Angeles, California 90025
United States
Tel: (310) 474-1938
Fax: (310) 474-2813

Roma Marble
7846 Sepulveda Boulevard
Van Nuys, California 91405
United States
Tel: (818) 781-6369

Eugene Salerno
916 First Street
Hermosa Beach, California 90254
United States
Tel: (310) 376-9097

Scantlin Custom Builders
P.O. Box 1022
Kerrville, Texas 78029
United States
Tel: (210) 896-5260

J. Robert Scott & Associates, Inc.
8727 Melrose Avenue
Los Angeles, California 90069-5086
United States
Tel: (310) 659-4910
Fax: (310) 659-4994

Sculpture Design Imports, Inc.
416 South Robertson Boulevard
Los Angeles, California 90048
United States
Tel: (310) 858-8266
Fax: (310) 271-9974

Sfreddo & Delgalo
392 Bridge Road
Pretoria North
Pretoria
South Africa
Tel: (27) 12 546 6067
Fax: (27) 12 546 6067

Southampton Brick & Tile
North Highway
Southampton, New York 11968
United States
Tel: (516) 283-8088
Fax: (516) 283-8349

Stoker Construction Inc.
57370 29 Palms Highway
Yucca Valley, California 92284
United States
Tel: (619) 365-8691
Fax: (619) 365-7189

Studio Marble, Inc.
11180 Penrose Street
Sun Valley, California 91352
United States
Tel: (818) 767-3436

Ray Stutzke
P.O. Box 137
Kootenai
Idaho 83840
United States
Tel: (208) 263-3975

Syndesis, Inc.
2908 Colorado Avenue
Santa Monica, California 90404
United States
Tel: (310) 829-9932
Fax: (310) 829-5641

Taung Marble
14 Kramer Road
Kramerville, Sandton
Johannesburg
South Africa
Tel: (27) 11 444 1290
Fax: (27) 11 444 1989

Michael Taylor Designs, Inc.
2415 17th Street
San Francisco, California 94110
United States
Tel: (415) 558-9940
Fax: (415) 558-9770

Terrajal
Monte Morelos - #550
Col. Alamo Oriente
Tlaquepaque,
Guadalajara
Jalisco
Mexico
Tel: (523) 657-7494
Fax: (523) 657-6726

Vetricolor
36041 Alte
Vicenza
Italy
Tel: (39) 4 44 696758
Fax: (39) 4 44 696174

Village Floors
1622 North East 205th Terrace
North Miami Beach, Florida 33179
United States
Tel: (305) 652-5104
Fax: (305) 653-4908

Walker/Zanger
8901 Bradley Avenue
Sun Valley, California 91352
United States
Tel: (818) 504-0235
Fax: (818) 504-2057

Russell White Special Finishes
16852 Septo Street
Granada Hills, California 91343
United States
Tel: (818) 360-2646

PHOTOGRAPHERS

Jerome Adamstein
153 South Robertson Boulevard
Los Angeles, California 90048
United States
Tel: (310) 289-8300
Fax: (310) 288-3885

Jaime Ardiles-Arce
730 Fifth Avenue
New York, New York 10019
United States
Tel: (212) 333-8779
Fax: (212) 593-2070

Victor Benitez
Guanajuato - #130
Col. Roma
Mexico, D.F.
Tel: (525) 574 8032
Fax: (525) 584 7571

Tom Bonner
1201 Abbot Kinney
Venice, California 90251
United States
Tel: (310) 396-7125

Dick Busher
7042 20th Place, N.E.
Seattle, Washington 98115
United States
Tel: (206) 523-1426

Grey Crawford
1714 Lyndon Street
South Pasadena, California 91030
Tel: (213) 413-4299

Billy Cunningham
26 St. Mark's Place, Apt. 4FW
New York, New York 10003
United States
Tel: (212) 677-4904

Francisco de la Fuente
Arapiles 17
Madrid 28015
Spain
Tel: (34) 1 445 8371

Jon Elliott
329 West 85th Street
New York, New York 10024
United States
Tel: (212) 362-0809

Phillip H. Ennis
98 Smith Street
Freeport, New York 11520
United States
Tel: (516) 379-4273

Eitan Feinholz
Atlaltunco - #99
Tecamachalco
Mexico 53970
Tel: (525) 294 7889
Fax: (525) 294 8513

Dan Forer
1970 North East 149th Street
North Miami, Florida 33181
United States
Tel: (305) 949-3131
Fax: (305) 949-3701

Jeff Goldberg/ESTO
ESTO Photographics
222 Valley Place
Mamaroneck, New York 10543
United States
Tel: (914) 698-4060
Fax: (914) 698-1033

Janos Grapow
Via Monti Parioli 21/Q
Rome
Italy
Tel: 06 32 44 831

Barry J. Grossman
18224 South West 4th Court
Pembroke Pines, Florida 33029
United States
Tel: (305) 433-5999
Fax: (305) 433-5999

Hickey-Robertson
1318 Sul Ross
Houston, Texas 77006
United States
Tel: (713) 522-7258

Lockwood Hoehl
55-52 Beacon Street
Pittsburgh, Pennsylvania 15217
United States
Tel/Fax: (412) 421-8285

Patrick House
P.O. Box 3759
South Pasadena, California 91031
United States
Tel: (818) 355-0183
Fax: (818) 441-5629

Timo Kauppila
Indav
Itälahdenkatu 18C
00210 Helsinki
Finland
Tel: (358) 0 700 14 31
Fax: (358) 0 677 149

William Kildow
1743 West Cornelia Avenue
Chicago, Illinois 60657
United States
Tel: (312) 248-9159
Fax: (312) 248-0227

Klein & Wilson
7015 San Mateo Boulevard
Dallas, Texas 75223
United States
Tel/Fax: (214) 328-8627

Bud Lammers
Advertising Photography
211 East Columbine, Suite A
Santa Ana, California 92707-9930
United States
Tel: (714) 546-4441
Fax: (714) 546-4813

Leland Y. Lee
201 North Orange Grove Boulevard
Pasadena, California 91103
United States
Tel: (818) 568-0919

David Livingston
483 Greenwich Street
San Francisco, California 94133
Tel: (415) 392-2465
Fax: (415) 398-3312

Scott McDonald
Hedrich-Blessing
11 West Illinois Street
Chicago, Illinois 60610
United States
Tel: (312) 321-1151
Fax: (312) 321-1165

Chas McGrath
3735 Kansas Drive
Santa Rosa, California 95405
United States
Tel: (707) 545-5853
Fax: (707) 545-0359

Norman McGrath
164 West 79th Street
New York, New York 10024
Tel: (212) 799-6422
Fax: (212) 799-1285

Jon Miller
Hedrich-Blessing, Ltd.
11 West Illinois Street
Chicago, Illinois 60610
United States
Tel: (312) 321-1151
Fax: (312) 321-1165

Michael Moran
245 Mulberry Street, #14
New York, New York 10012
Tel: (212) 226-2596
Fax: (212) 226-2596

Mary E. Nichols
132 South Beachwood Drive
Los Angeles, California 90004
United States
Tel: (213) 935-3080
Fax: (213) 935-9788

Anthony Peres
645 Oxford Avenue
Venice, California 90291
United States
Tel/Fax: (310) 821-1984

Michael Portman
The Portman Companies
231 Peachtree Street, Suite 200
Atlanta, Georgia 30303
Tel: (404) 614-5089
Fax: (404) 614-5555

Lanny Provo
120 North East 100 Street
Miami Shores, Florida 33138
United States
Tel: (305) 756-0136

Tuca Reinés
Rua Emanuel Kant, 58
São Paulo, São Paulo 045536
Brazil
Tel/Fax: (55) 11 852 8735

Kim Sargent
1235 U.S. Highway One
Juno Beach, Florida 33408
United States
Tel: (407) 627-4711
Fax: (407) 694-9078

Jordi Sarra
Zamora, 91-95
08018 Barcelona
Spain
Tel: 485 04 00
Fax: 485 07 42

Carsten Schael
1A Yalford Building
52-58 Tanner Road
North Point, Hong Kong
Tel: (852) 565 6349
Fax: (852) 565 6926
Pager: 111-8363 A/C 811

Bob Shimer
Hedrich-Blessing
11 West Illinois Street
Chicago, Illinois 60610
United States
Tel: (312) 321-1151
Fax: (312) 321-1165

Julius Shulman, Hon. AIA
7875 Woodrow Wilson Drive
Los Angeles, California 90046
United States
Tel: (213) 654-0877

Paul Stowers
25560 Soquel - San Jose Road
Los Gatos, California 95030
United States
Tel: (408) 353-8947
Fax: (408) 247-1242

Julio Tovar
Priv. Agustin Melgar
#1706 NTE.
Monterrey N.L.
Mexico
Tel: (528) 374 29 98
Fax: (528) 374 58 80

Jennifer Tull
4416 East Calle Feliz
Phoenix, Arizona 85018
United States
Tel: (602) 852-0396

Rafael Vargas
Gran Via Corts Catalanes, 699
08013 Barcelona
Spain
Tel: (34) 3 232 6727
Fax: (34) 3 232 4151

Alex Vertikoff Photography
1009 Vernon Avenue
Venice, California 90291
United States
Tel: (310) 450-9442

Peter Vitale
P.O. Box 10126
Santa Fe, New Mexico 87504
United States
Tel: (505) 988-2558   (212) 888-6409
Fax: (212) 838-7369

Paul Warchol
Paul Warchol Photography
133 Mulberry Street
New York, New York 10013
United States
Tel: (212) 431-3461
Fax: (212) 274-1953

Alan Weintraub
2325 Third Street, Suite 325A
San Francisco, California 94107
United States
Tel: (415) 553-8191
Fax: (415) 553-8192

Toshi Yoshimi
4030 Camero Avenue
Los Angeles, California 90027
United States
Tel: (213) 660-9043
Fax: (213) 660-2497

INDEX

ARCHITECTS AND INTERIOR DESIGNERS

Manufacturers, Distributors, Suppliers, Contractors and Artisans

PHOTOGRAPHERS

ACKNOWLEDGMENTS

Already having benefitted twice from an enlightened working relationship with PBC International and the high standards established by Publisher Mark Serchuck and Managing Director Penny Sibal, I am thrilled to have the opportunity to create a series of four more books. That Managing Editor Susan Kapsis has overseen and scrutinized their development fills me with a sense of security. Besides, with our interests being similar and our enthusiasm high, we have had a marvelous time!

PBC's Technical Director Richard Liu has again lent his expert analysis to make sure that only excellent photographic material prevails. The art department's Barbara Ann Cast proved indispensable to the final execution of the book's layout. And to members of the editorial department—Debra Harding, Francine Hornberger, Jami Hall and Christine Brako—a million bouquets! Many thanks to Anistatia Miller for the book's sensitive, accomplished and intelligent design.

I was also most grateful to have had James Gabrie, Beatrice Jakots and Tuula Stark come to my aid when I was desperately in need of translators, and to have had the admirably thorough Angeline Vogl proofread every word.

The list grows as I try to list the many people who have supported this endeavor as I have called their offices across the United States and throughout the world. Among those who have been especially helpful are:

Gloria Blake, Ph.D., *Haut Decor* Magazine
Judith B. Gura, The Gura Agency
Robert Hund, Marble Institute of America
Päivi Jantunen, Hackman Designor/Iittala Glass
Robert J. Kleinhans, Tile Council of America, Inc.
Gray LaFortune, Ceramic Tile Institute of America
Mary Levine, Tile Promotion Board
Patti Richards, FIIDA, Residential and Contract Interiors
Judi Skalsky, Marketing Consultant
Brian Trimble, Senior Engineer, Brick Institute of America